The Wondrous Journey of Food

By
Ronald R. Van Stockum Jr.

Illustrations by
Steven P. Eilers

THE WONDROUS JOURNEY OF FOOD
By Ronald R. Van Stockum, Jr.

ISBN: Softcover 978-0-9974242-5-6
ISBN: Hardcover 978-0-9974242-6-3

First Printing June 2017

Bareham Tsuga Press
4163 Zaring Mill Rd.
Shelbyville, KY 40065
rvs@vanstockum.com

Printed in the United States of America

The Wondrous Journey of Food

Table of Contents

EILERS

I.
A Necessary Journey

I am lost. Somewhere in the world, I am lost. Is it in that jungle of forest in Eastern North America? I suspect so. But not of modern time, not my time. Maybe long ago? Or perhaps sometime in the future?[1] What has happened?

From the constellations at sunset, I surmise it must be late summer.[2] Tea is steaming above the scorpion's tail, and trees are modeling their fall fruit adornment. Leaf brittle trails release a musty scent as I pass.[3]

I am clothed and cloaked and alone. And I am hungry.

My thirst is slaked, however, for there is an abundance of water here.[4] But it is wild water, full of living forms consciously removed or killed from the previous water supplies of my people.[5] But my people are gone, or I have left them somehow, for there are no treatment plants here. No buildings of any sort. But I care not now. I am hungry. And those living things now living in my gut are hungry too.[6]

Where should I look? Where would you look?

There, a snake. What kind? I have to act quickly as it is sidling away. Smelled me before I him.

I begin the chase, quickly going over the characteristics of a poisonous snake.[7] Probably a pit viper. A bounding leap brings my boot down on its head.

Good, I still had my boots. I fumbled in my pockets. No wallet, no money. Who cares! I am hungry. There, in my right pocket, is my small penknife dangling from a tiny flashlight.[8] Great, it works. But this blade is too small and would be awkward in use to dispatch the snake.

EILERS

I carefully bend down, keeping pressure on the snake's head. I will need to examine that head to make the snake identification, although I can tell by the arrangement of its anal scutes if the darn thing will just stop squirming.[9]

There, I have it! I pinch my thumb and forefinger and tightly squeeze just below its head. As I raise up the snake, it writhes wildly, lashing and wrapping its body around my forearm. I bring its head up to my face for a closer view.

Round eyes - not slits like cat eyes - fearfully looking at me. There is the nose slit and I see no heat-sensing pit between the eye and the nose. Non-poisonous.

The snake is almost 20 inches long, and thin. Not much of a meal, I think. I can identify its species, but will not identify it to you. I am somewhat displeased at what I am going to do next.

Grasping its tail a third of the way up with my free hand, I release the snake and swing it furiously against a tree. Over and over I swing, smashing it against the tree until I am certain that I have crushed all life from its small head. I am very hungry.

I examine its now limp length. I run my fingers along its thin form. Slim pickings, I think, literally so. I could filet it with my little penknife, but my stomach wretches at the thought of chewing through those slippery, tawny fibers of muscle. Would they twitch in my mouth?

I decide to cook it first. Matches?

I fumble through my pants and coat pockets. None. In fact, nothing at all. How did I end up in this place with only a penknife and tiny flashlight?

Now how was I going to make a fire to cook that snake meat?[10] Come to think of it, a little warmth would be good, too. And maybe there are animals out here hunting me for food just like I am hunting snakes. I spin around quickly at a sound behind me. Nothing. A shudder runs across my back. Yes, fire would be good.[11]

My options are limited but not as sparse as those available to my ancient ancestors. They had probably first captured fire from lightning-induced grass flames. Some may have even learned to spin a reed between their hands or in a bowstring and created those first friction fires on a piece of wood. And surely, after learning to exploit flint rock for weapons and tools, they noted the sparks when that rock was struck.[12] I have an advantage. I have a steel knife blade. It is small, but large enough to strike flint against.[13]

Yes, flint will work on my blade. Now where to find it?[14]

Caves? Flint nodules can be found in limestone and I do have the tiny light. No, that light is a most limited tool. Useless after the battery is exhausted. I shall defend its light zealously.

Maybe other people came here before me. Maybe many people have found themselves mysteriously deposited here to live out a primitive life and expire. Maybe they made tools and used flint. And maybe some of those tools washed and rolled into the streams draining this valley. I will need to find a stream-cut bank to look at the exposed layers of soil. Maybe, just maybe, I will get lucky and find flint rock washed down and now exposed in the soil column sliced open by the stream.

Downslope, I move. The dead snake is slung over my shoulder. I am still hungry. Here I go. There is a floodway adjacent to the creek.[15] There should be good pickings in the opposite stream bank.

But wait - what is this? The grasses in this meadow are unnaturally aligned. And they are similar in appearance. Have they been planted? Some in primitive rows, some in mounds. People did this!

EILERS

I crouch low in the adjacent patch of tall prairie grass and look around for danger. Nothing moving. I wait a long time, listening for the return of the gardeners. There must be other humans, banished here like me.

They too would have been hungry. So hungry that they had desperately turned to growing plants with tiny seeds for nourishment. They will never realize that those tiny seeds will raise up great civilizations with even greater numbers of people.[16] And those people will come to reproduce beyond the carrying capacity of the land. But that will come much later. For now there would be a more primitive competition for calories.[17]

It would be easy to see them killing me, an unwelcome resource invader. Killing me like a snake. Would they cook and eat me, too? Lots of calories, and all of the essential amino and fatty acids.[18]

That cold shudder returns to my body, but this time runs down the sinewy tendons along my spine. Tendons that they would gently strip off my back, soak in water, and bind wooden shafts to their stone tools and weapons.[19]

Yikes!

I got up. Hunger made me do it. I headed to the creek bed. My path led directly through the garden. I hesitated, but for a very short time, just long enough to recognize the species cultivated. That told me of their nutritional needs.

As I passed through the garden – that term is not really accurate by our definition – I was able to catalogue the plants. College does that to its students, exposes them to science and the arts of which they are otherwise ignorant. College raises the lid on the treasure chest of knowledge. I had lifted that lid and feasted on the glowing jewels within.

But I would have spent too much time in college, learning and teaching. Delightful. Eventually I had to choose a degree and leave. So biology it was, and law for a living. I wonder what would have happened if I had chosen English? Hah! A better storyline perhaps! Tighter syntax, better use of phrases, and greater exploitation of rhetorical devices. But as spontaneous?

I had been exposed to the plant kingdom and I was aware of the historical development of these primitive gardens and their contents. Here were tall goosefoot, and the even taller marsh elder. And off to one side, in an area burned clean by fire, grew numerous clumps of little barley, its clusters of tiny seeds bouncing atop its short stems as though fruit baskets waiting for the plucking. I looked around further. I knew what this assemblage was. Words have power and these plants were part of an assemblage we call the "Eastern Agricultural Complex" of the early Native Americans.[20]

Now where were they growing sunflower? Those bright yellow petals with the dark dotted center of seeds. Surely, with such a loud display, they would announce themselves. I was caught up in the hunt. Again, the search for knowledge. So captivating is the effect of learning. I had forgotten my concern about the return of the gardeners!

I probed the edge of the field near the creek. The dead snake body brushed off my shoulder. It fell lifelessly across my shoe. As I reached down and picked up its limp body, I again felt both hunger and revulsion. I tucked its body under my belt.

Need to find flint, I remembered. Into the woods by the stream I went.

Bingo.

Sunflower!

EILERS

No, wait. This flower is smaller, growing in a copse of sunlight penetrating to the stream bank.

It's the Earth Apple! How did it ever come to be called the Jerusalem Artichoke?[21] I grasped one of its stout stems - tugged, twisted, and pulled. Pop! Out came its roots and tubers. Those beautiful, swollen, maggot-looking tubers. My mouth squirted shots of starch-breaking enzymes onto my tongue. Oh, I was hungry.

But I knew better. The sugar in this plant's storage center was in the form of inulin, a polysaccharide of fructose for which humans have no enzyme sufficient to break out the sugar. Left to ripen, these tubers break down into sugars naturally. Fructose. Very tasty. I wondered if cooking accomplished the same thing.

The stream was dry and I hopped down onto the flat limestone rock bed of the watercourse.[22] I could see several banks of exposed soil cut by the watercourse in its meanderings.

I squatted, careful not to step on my future snake meal dangling from my belt. I could see round rocks and smaller pebbles protruding from the opposite cut bank. None appeared like the unnaturally "flaked" surfaces of rocks worked by Native Americans. I stepped over the creek bed and crossed to the other bank. As I probed the clay with a stick, I thought of what else I hadn't seen. Corn.[23]

If I could date the time period within which I now found myself by agricultural practices, I might obtain a more certain knowledge of my predicament. I saw no corn being grown in that garden. No mounds of soil with bean trailers crawling up a cornstalk and squash vines smothering out weeds competing at its base and acting to retain moisture. That practice involved the "Three Sisters." Those Sisters were Mesoamerican cultivars that spread into North America after more temperate hardy corn strains were developed. And that didn't arrive here until about 1,500 years ago. So, if I was right, and these were Native American gardens, I was now living in what our anthropologists call the Middle Woodland Period. There I go. Convenient phrases used to encapsulate loose knowledge. Freud again.[24]

But I was really hungry. I thought of taking a bite out of the raw snake.

Yee Gads! What of its intestines? And would it bleed into my mouth? On to the next clay bank I went.

I had better luck there. Infrequent slivers of flint flakes were embedded in the bank. Someone upstream had been knapping a core ball of flint, making points, edges, and tools. If I could only find one of those larger flint cores, I would be able to hold it tight enough to strike sparks off with my knife blade.

There you are! I see a knapped edge protruding from the clay. I pull it out, wiggling it as I do. It is gorgeous. A beautifully worked flint knife. The tip is broken off, but the edges are still sharp. I drag them across my palm. Ouch! Very sharp. A new tool. I pocket it and continue searching with excitement.

Night is coming. I don't like that. I jump onto the bank and start making a small pile of tinder for a fire. Strips of bark, small twigs. Out with my tiny blade. Strike against the flint. Wow, this is not working. Hard to call these things sparks. I am not going to start a fire this way.[25]

Becoming a little desperate, and perhaps somewhat foolish, I consider trying to create a friction fire by spinning a small shaft of wood between my palms. I certainly do not have the patience to succeed at that practice, nor had I ever learned the elements of such a fire-starting technique. I look around in the dimming light, and then again at the dead snake in my belt.

EILERS

And there, beyond the snake, I see it. A rattlesnake fern! It is funny how mental association works. That fern might be of some help, but it gets me thinking of other primitive plants; ferns, fungi, and algae. Then it hits me - lichens! Dry tree lichens!

I quickly scout around, careful to look on the north side of tree trunks - at right angles to the setting sun. They were there, of course. Algae, cyanobacteria, and fungal symbionts living entirely in self-sufficient lichen harmony. I wonder if I can eat them too?

No time. I finger the fruticose and foliose forms and tear off a handful of dry, crusty growths. Wasting no time, I hold them against one side of the flint core and begin to furiously strike my steel blade against its surfaces. Sparks are sent into the vegetative mass, which flares up and then quickly dies out.

Back to the rattlesnake fern. I get down and look closely at its short pinnately dissected frond. You might think it was a young Poison Hemlock plant but for the realization that it will grow no taller. And the Poison Hemlock is an European import anyway. Socrates, etc. Not here yet.

What I want now are the fern's tight collection of sporangia on an upright fertile frond. Dry, oil-rich spores. The stuff of magicians, witches, and wizards. Throw a handful in a fire and, pop, a disguising column of smoke will rise.

I run the stalk of spores through my fingers, collecting a small group in my palm. Then, carefully, I sprinkle them within the crevices of the lichens held against the rock with my other hand.

Out again with my penknife. Scratch, scratch, strike, strike.

Pop! The spores begin to catch hold of the sparks.

EILERS

I began striking and blowing. In a short time, using longer, softer breaths, I had a promising flame in the bed of dry lichens. I gently laid the pile in the tinder of the fire pit that I had built. As darkness filled the spaces within the streamside forest, I lay back against a fallen log, roasting my fileted sliver of snake meat.

It was delicious. Tasted like chicken. But, of course, there are no chickens here.[26] They would first have to be brought from Southeast Asia, where they developed, into Europe and then here, as part of the Columbian Exchange. Same for the Poison Hemlock.

I was not cold that night. My jacket was warmed by the fire and, like our planet, reradiated heat into my sleeping body.

Yes, I slept. You would have too, if you had been through what I had. Tummy full, my mesentery was feeding. Not enough blood for both digestion and muscular activity. Sleep came without my awareness of it.

The next morning, however, I was hungry again. Snake meat is good for stocking up on amino acids, but I now wanted base energy - the explosive kind found in sugars and carbohydrates, or the richer load of fat and oil stored for life's energy needs.[27] I could taste the sugars dissolving in my mouth.

I warmed up by the fire and, after the sun rose, heated up in its microwave radiation as I headed back out to the meadow.[28] I added some wood to the fire. I didn't want to go through the difficulty of fire-starting again. I had "captured" my fire and I intended to keep it.

As I rose up on the small hillock, I found the field covered in beautiful white snakeroot. It is interesting to note that when one is hungry in the woods, with no hope of home and safety or cupboard and icebox, plants look different to you. Deliciously different. And I could have just reached down and grasped a handful of the snakeroot's white flowers, rolled them up in one of the plant's green spatulate leaves and chewed on it like a fat, soft sausage.

But I didn't.

Some things used to be taught in families. The danger of white snakeroot is still taught by farmers of my time, especially at dairies. But there were no dairies here – farms neither. I had learned the danger of that plant in class.[29]

If plants look nutritious in the wild, then slow fat animals look even more so. No more tiny oil-rich seeds eaten to obtain those energy-rich fatty acids. Animals like the acorn-plump squirrel and chubby round groundhog look like perfect meals. Lots of extra energy to store up, and surely those two fatty acids that we cannot manufacture ourselves. Ones we must get from our diet. And, of course, in the animals' muscles we would find the twenty amino acids that we use in our body's structure and metabolism, including the nine that must also be obtained in our diet.

I kept walking as I thought. Thinking of food. That is all you do when you are hungry. Fantasize about it.[30]

Whoa! Huge birds sluggishly lifted off and hovered above me. Bald heads. Vultures! Carrion! Fatty meat!

I quickly crossed the hillcrest and saw it. A dead deer, deflated it appeared, by the birds tearing at its innards. Lovely birds. They clean up the mess. We protect them so.[31] But away now, it was my turn. I raced over to the carcass.

Yup, plenty of muscle and fatty tissue left. A gold mine! I sat down and tore off some tissue. The odor of that spot was pretty rank, but I factored in that the animal had been dead for a while and moved forward in my mission. I clamped down on a dry, greasy piece of meat and jerked it around until it tore off in my mouth.

EILERS

Yuk! Yee gads! It tasted slimy and rancid and sour at the sides of my mouth.

Bad! I thrust it out of my mouth and rose up spitting.

It was disgusting and now the odor made me want to retch.[32]

I headed back downslope to the good air. Plant food it would have to be. Though I did wonder in passing if I might snare one of those fat squirrels. The flint blade that I pulled from the bank would skin them quickly.

I shift again to the thought of carbohydrates and the enzyme amylase squirts from the sides of my mouth. Sugar, that's what I want. Maybe some greens, too. I look around. Maybe some dandelions or chicory greens. I love their yellow and blue flowers.

But there are none, of course. Wherever I was, I at least knew *when* it wasn't. It wasn't after Columbus "discovered" America. Dandelions and chicory plants are not native to the New World. They had probably been brought in as hitchhikers. Travelers within the Colombian Exchange.[33] And they were not here now. Day lily bulbs? Same thing. European in origin. None here.

Nope, definitely not modern times.

Back to the stream. I walk among the floodplain flora and see the carpet-like expanse of mayapples. Maybe there are some of those singular, dangling, red-ripe fruits hanging from the green parasol of leaves spreading out on top.

I know, I know, they must be completely ripe or they are toxic. The fruits are gone anyway. Gone in a mid summer ripeness. I want to bite into one of those meaty-looking leaves brushing against my face as I look underneath. Just like a box turtle, I want to nibble away. Hunger makes you like an animal, but I had learned not to eat that leaf. School again.

Maybe there were some Paw Paws within the understory of this streamside forest. Paw Paws are also known as Kentucky Bananas! Maybe there is a persimmon tree. No, better to wait for winter for those fruits. They need to be soft and ripe to avoid that bitter taste. More squirting mouth juice. So sweet and tasty.[34]

Look, a new puffball. I reach down and squeeze it. Fresh. Don't eat the mushrooms, I say to myself. You don't know them well enough to avoid the poisonous ones, and many species look alike.

Dagnabit! I was hungry. In frustration, I grabbed the leaves of one of the greenish herbs nearby and stuffed them in my mouth.

Ugh! My mouth twisted up like roasted chestnut skin. My tongue withdrew into my throat, and I gagged as I expelled the bitter mass that swelled up with my forsaken saliva.

What toxic defense had that plant raised against ingestion? It could have warned me like the monarch butterfly. I knew better than to eat that lepidopteron![35]

I went to a pool in the creek to wash and cleanse my palate. Minnows! Now what would it take to eat those little beasties?[36] Saliva ran again as I thought of plopping that piscine into my mouth. I was over the squiggle fear. I was hungry. One bite with my teeth and a swallow. That's all it would take. A delicious thought.

Back to the stream bank. I pulled down a sapling. Its spring pulled me up and over the root-bound grass clumps bordering the stream. It was easy to traverse the riverside hardwoods swept clean of underbrush by frequent freshets.

I climbed up the slope behind the creek. Now, how did those Kentucky Bananas get up there? I saw the distinctly ovate understory leaves upslope. I know of no native animals with large enough guts to allow passage of their gigantic seeds.

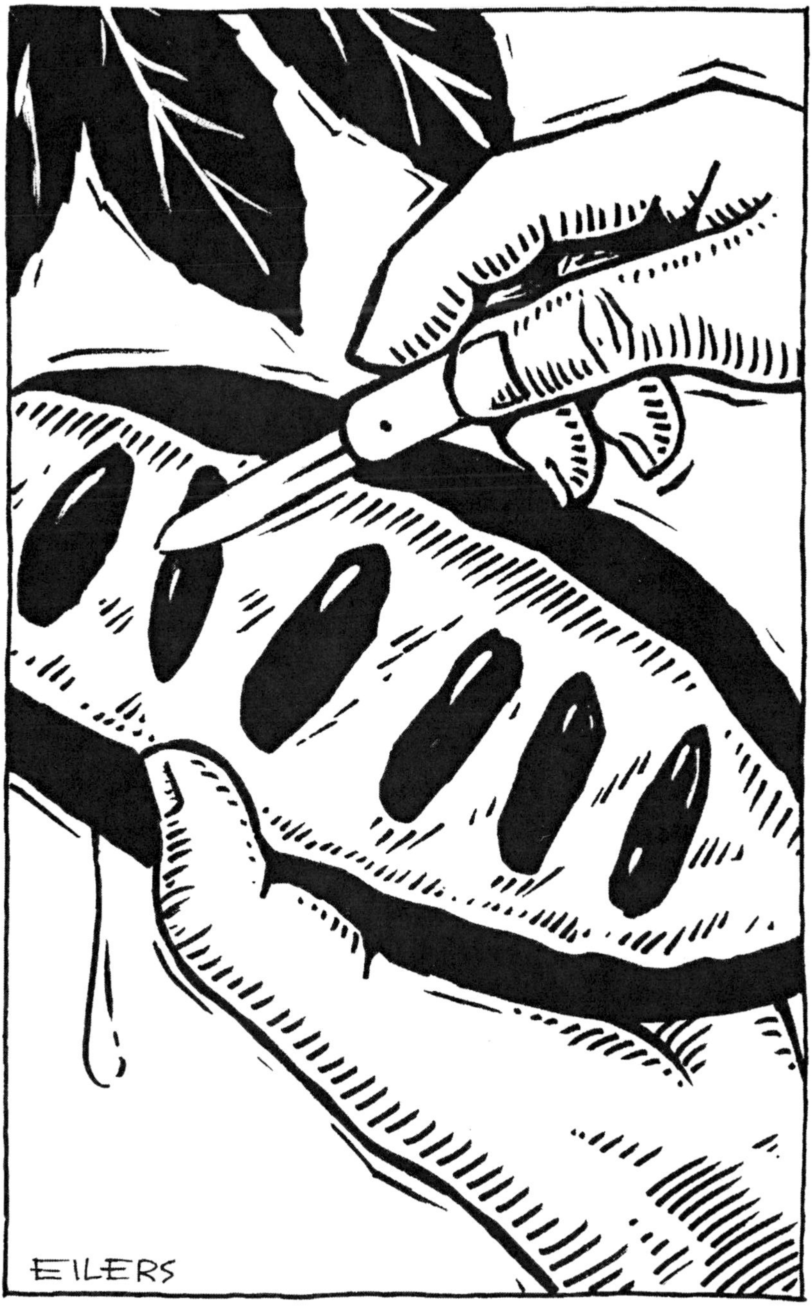
EILERS

No matter. I could see the thick, oblong fruits. I was in time. The pulp thieves, animals like me who would feed on that exotic, aromatic custard flavored fruit, had yet to tear them apart.[37]

The mind is a great thing, and imagination even greater. A flood spilled out of the corner of my mouth as I reached the understory cathedral of Paw Paw clones that represented this forest, all linked together as one individual.

I grabbed a low-hanging fruit. I squeezed and snapped and soon had one in the palm of my hand. My penknife quickly sliced it open. I pried out some of its yellowish flesh and swept it into my waiting, open mouth. Oh, how wonderful. Is this how the Pharaoh felt when he tasted the first orange? I kept scooping, but it immediately became obvious that the seeds took up almost all of the room inside the casing. I grasped one and it slipped between my fingers. It was almost the size of a half dollar and swollen in width like a thick black tick. I rolled it around my teeth, scraping off the titillatingly tasty, aromatic pulp.

I was fearful of catching the seed in my throat. My God, how big were the animals that evolved to ingest and disperse these seeds?

I ravaged five of the fruits right there, and pocketed more that I could reach. Those remaining would not last long on the tree, but I would return the next day and collect again. I headed back to my campsite and its fire.

Just in time, too. I raced over when I saw no flames. Just embers.

I quickly knelt before that religious portent and gently blew life into it until flames returned. Then I slowly placed dried leaves, one at a time, blowing each one aflame until I felt safe enough to leave it and get more twigs and small branches. Eventually, a teepee of dried logs surmounted the tiny basket of

EILERS

tinder barking out shouts of tiny flames. Soon enough my log temple was all aflame in my honor. I curled up near it, relaxing in the carbohydrate metabolism of Paw Paw pulp.

It does not take your digestive enzymes long to dispatch the various carbohydrate packages into glucose. And it mostly happens in the small intestine, thereby quickly emptying out the stomach.[38] There I go. Hungry again!

I remembered the logs that I brought to the fire. One of them had a peanut-sized grub under it. A dull, whitish, plump accordion with a bushy head and busy mouthpiece.

Come to think of it, the groundnut is like a peanut and surely grew around here. I saw the winding vines and pinnate leaves with lancelet leaflets in my mind. And also in the vine category, did I see any wild grapes? I would need to be careful with those and not eat a similar-looking moonseed. Its leaves are like those of grape but its delicious-looking berries are very toxic. I pictured both in my imagination for later investigation.[39] I was off after that grub.

I reached the spot where I had removed the log and brushed aside the leaf mull. There he was. Wriggling like he knew dinner was on my plate. I gently lay down sideways on the forest floor and reached over to pluck out the beast. I grabbed it by its fat little body, which began testy movements, writhing at my touch. I knew what I would do, even though I had never done it before.

I inserted that little black head between my front teeth, snapped off the primitive cerebral center, and spit it out before it could grasp onto my tongue with its mandibles. It couldn't, of course. It had no significant biting parts. It was a grub, after all. But I was not going to dab its head with my tongue tip to verify that fact.

I was quick to return to that still-squirming, fat little body. I placed its headless shoulders to my lips and half squeezed,

EILERS

half sucked the colloidal suspension of fluids that flowed into my mouth. I threw down the exhausted casing and licked my lips.

Now, where had I learned to do that? I think hunger remembers more than we do. Anyway, the slime tasted smooth, savory. Umami! Glutamic acids and nucleotides, no doubt.

But it was rather bland tasting. I looked around for something to spice up my meal. The adjacent prairie meadow seemed promising. There had been no fields of waving yellow mustards this fall, I knew. European plants again. But I had seen some native species up there.[40] Soon I was chewing a pepper-green mash of leaves, cutting into and wiping up the grub mixture of intestines, organs, muscle, and slime from my mouth. My stomach had shrunk some after the Paw Paw orgy. It would take more time to digest the grub paste.

Satisfied, I put more wood on the fire. That would be a primary occupation of mine in the days to come. I began to consider hunting. Killing animals. That would take tools and weapons. I thought back on my archaeological studies. Rocks and sharpened sticks. Spears with ligament sinew tying on flint points. The incredible rapid-fire weapon of the Native Americans, the "Atlatl."[41]

I drifted off into a warm nap, wondering if I would be there long enough to crossbreed some of the plants in the garden.

But those plants had already been selectively bred. And well hidden behind the bank and the brush were those planters who were determining my future.

Snake meat or adopted tribal member? The planters had not yet decided my fate.

From behind the stream bank, one of the warriors rose up quietly and cocked his spear thrower directly to the back of

his shoulder. He then let the dart fly, the swoosh o
movement muffled by the winged bannerstone weight on th
throwing shaft.[42]

Accordingly, I did not hear of its passage until it collided with its target.

II.
Author's Note

So, my dear readers, I choose to begin this study of food in a short work of fiction. An adventure story, yes, but one rooted in fact.

I hope this unconventional beginning will catch your attention as you, my protagonist, progress through the time sequence of culture, food, and regulation that I present in this work.

I endeavor to write in an engaging, informative style, making use of grub-like footnotes hidden beneath the text and full of savory detail. That is where you will find the supporting science, law, and society that feed this work.

This exercise in educational exploration moves forward as an intact, entire vehicle within the text alone. Therein lies the adventure in learning of this fascinating subject. Storyline and data too!

The footnotes may be laborious, but I hope rewarding each unto itself. Few readers will perhaps exhaust those references, so I feel safe in indulging my interests. For example, the lack of a magnetic field on Mars is a sobering reality for those seeking Elysian Fields for the expansion of humankind on that planet (see footnote 11). Likewise, a discussion of the Milankovitch Cycles might upset your standing on this "Terra Firma" (see footnote 84). What would you eat if you were stranded on a tropical island for 28 years (see footnote 30)? Maybe the unthinkable (see footnote 52). And maybe you will miss the humor that I am otherwise able to bury within the footnotes.

But the story itself is a good read, I think. For those who read it, pleasant steps. For those more adventurous souls who will sail within the footnotes, fair wind. I will see you in class!

III.
Introduction

You would think that your experience with a variety of foodstuffs is vast. So you might consider that describing that from which you draw your daily enjoyment and sustenance would be a relatively simple topic to set out. Outside of exotic foods and unusual dishes of ordinary foods in our diet, there do not seem to be so many different foodstuffs in our daily lives. And within them all there is only a small subset of organic chemicals in four general categories.[43] Of the 118 natural elements now listed on the periodic table, only a handful is utilized by the human body in any great amount.[44]

Yet when you think of food, it is probable that you do not think of it like an organic chemist. You don't analyze it in a way that a biochemistry student might explore the backbones of carbon atoms peppered with hydrogen highlights that make up the great variety of life's "hydrocarbon" products. If you did, you might muse, momentarily, about the wondrous diversity of sugars and carbohydrates produced in the varied plant life on the planet.[45] You might even hesitate long enough to consider that almost all of life relies on the creation, consumption, and metabolism of one simple sugar, glucose - six carbons, twelve hydrogens, and six oxygens.

And in your assessment of food, it will not take long for your personal history with the subject to surface. It is a subject, the memory of which is triggered by the many tastes, hungers, and odors that you have picked up from your living environment. Original survivor skills.

The sense of smell is itself a wondrous ability of humans to catalogue the great diversity of organic compounds in nature, some food and some poison. But we will not begin the history of food in civilization through smell. We will start with the examination of taste. Taste does, after all, more immediately drive what we ingest, and as we will see, why we eat. Thus, I begin this descriptive journey, hoping to elicit your own memories of taste, and perhaps even spice them up with long-held whiffs of familiar scents.

IV.
Taste

Humans are a species of the genus *Homo*. Carl Linnaeus gave us that name in 1758. We are members of the family *Hominidae* in the order of Primates. We are all designated as part of the class, *Mammalia*, in the phylum *Chordata* of the kingdom *Animalia*. Although as *H. sapiens*, we are the only living species of the genus *Homo* extant, there have been at least seven others before, and in some cases, with us. They have colorful names such as Handy Man, Java or Peking Man, the Hobbit. The most recent of our sister species was called Neanderthal Man, *Homo neanderthalensis*.

It turns out that *Homo sapiens*, modern humans as we like to call ourselves, interbred with the older species, Neanderthal Man. That species is now extinct, which may have had something to do with the competitiveness and violence of our species. But we can still track the presence of Neanderthal Man through his genetics. And our genetics. For you see, on average, modern humans contain between 1-4% of DNA that was contributed from the Neanderthal genome.[46] In fact, about 20% of the total Neanderthal genome survives in different populations of our people.

Neanderthals gave us varying skin and hair characteristics, some of which are found in more than 65% of our European population. Modern humans also received some deleterious genes from Neanderthals. Their contributions may impact the occurrence of Type II Diabetes, Crohn's Disease, Lupus, smoking and depression in our species.

But what did these ancient species of the genus *Homo* eat? There is a lot of colloquial fantasy, and some scientific investigation, on the subject. And surely much of what they ate was determined by taste.

We have identified five elements to the sense of taste, and they are not just pleasure principles.[47] There are even more elements to the sense of smell, and smell may be just as important as taste to our eating habits.[48] Indeed, when breathing through our nares, we are "tasting" the air. But back to our tongues.

Taste is flavor data, rewarding us for eating some foods and warning us away from others. We are all familiar, for example, with the power of sugar. Sweetness. Addicting. Our body bases its energy metabolism on just one type of hexose, the six-carbon simple sugar glucose. Other sugar units (there are sixteen six-carbon hexoses and fructose, a six-carbon ketose) are converted in the glucose oxidation cycle when we burn them for energy. We call that process respiration. Breathing. Oddly, glucose is not the sweetest tasting of these simple sugars. That is fructose, which is present in high quantities in honey and also in the indigestible polysaccharide inulin (think Jerusalem artichoke).

So we are addicted to sweets because we are addicted to energy and the plants that trap it and the animals that concentrate it and convert it to other important human building blocks in their bodies. When you are hungry, you can dream of sweetness and crave the taste. It is a survival mechanism.

The taste of sour is protective. It is the warning of decaying, rotting food. Such matter may contain dangerous pathogens, causing sickness and even death. Nevertheless, in the cuisine culture of certain societies, sour is intentionally added to enhance food dishes. And then there is the enticingly sweet, and very sour, lemon!

Bitter is another defensive taste. Many plants and animals produce toxins as a warning to predators. Another survival mechanism of the tongue. Stay away.

Salt is integral to the operation of our cells and movement of materials across its membranes. It also aids in preserving foods in longer storage, preventing them from rotting. The taste for salt is a powerful urge. A powerful need. But salt can be overwhelming. Try drinking saltwater. You will die of thirst cast adrift on an ocean.[49]

Perhaps the most curious taste of all is the fifth one, umami. Discovered and described just over 100 years ago in Japan, it detects a "savory" smooth, enticing flavor. Its discovery was the reason for the addition of monosodium glutamate (MSG) as a flavor enhancer to foods. It informs the palate of the presence of amino acids, many of which we cannot manufacture and must ingest.[50] Melted cheese is high in umami. Is that why we love pizza and cheeseburgers?

V.
The Paleo Diet

Then there is the question of what foods our distant forbearers digested. Were they meat eaters? Herbivores? Or were they just opportunistic feeders, consuming whatever was in season or that they could catch? Is it our teeth that give it away? Our digestive system? What did Neanderthal Man eat?

There are conflicting theories concerning our ancestors' diets. It is probable that they were omnivorous, eating the wide variety of foods available at different times of the year. And there is some evidence of possible cannibalism.[51]

We do know that grass grains and cereals had yet to be cultivated and animals had not yet become domesticated. Humans were foragers, gatherers beginning to develop tools for hunting. Roots, tubers, and fruits would have provided complex carbohydrates with honey, perhaps, providing the infrequent and exceptional taste of sweetness. But how would they get the necessary essential amino and fatty acids from their diets? Animal tissue was surely important. And genetic selection for tool and weapon-making traits accelerated the capture of game. But it may have been the taming of fire that provided the most important innovation in hominoid nutrition.

Fire, of course, provides light and warmth. It can also be a weapon. And it inaugurated a new occupation for our hominoid ancestors. Cooking. And cooking is not just for the benefit of taste. It is argued that cooking provides new, more easily obtained sources of nutrition, allowing for a surplus of calories to feed the evolution of larger brains. Cooking can also change the composition of foods available for absorption, releasing essential nutrients not otherwise available. It is also important in preparing mixtures of different foods that combine to provide essential nutrition. And boiling can leach out poisons, making available new sources of nutritious food previously thought too toxic to eat.[52]

But the biggest nutritional development of our species was the increase in the availability of calories created through agriculture. That development required, however, the counter intuitive realization that small grains and kernels, if gathered in sufficient quantity, can produce an excess of caloric needs for a family. Intentional gardening and selection of larger, more nutritious strains of plants produced even more calories per plant. The agrarian revolution had begun. Family structures changed, populations increased, and civilization, at least as we describe it, began. But some of the best soils for growing crops were arid and needed irrigation, and others flooding. That effort required coordination and group effort. Society and order. Bingo!

VI.
Civilized Beginnings

Hammurabi's stele was erected in Babylon almost 4,000 years ago. The stone stele sets out 282 laws carved in cuneiform, including the principle of "an eye for an eye" (Lex Talionis) and many provisions dealing with contract as well as domestic relations. The provisions of interest to us here deal with "corn." The problem is that "corn" is native to the Americas and was not introduced into the "Old World" until after Columbus' voyages to the "New World" in the 1400s A.D. What gives?

Here is an example of an early translation of the cuneiform on Hammurabi's stele; "If he give a cultivated cornfield or a cultivated sesame field, the corn or the sesame in the field shall belong to the owner of the field..." (The Avalon Project, Yale Law School; avalon.law.yale.edu). Even Austen Henry Layard uses the word "corn" in his description of Assyrian Bas-Relief sculptures excavated in Nineveh.

The explanation can be found in the Oxford English Dictionary. When Hammurabi's stele was translated into English in the early 1900s, the translator used an old agricultural term for whatever grain was being grown in the field. That word in England's English was, and still is, "corn." But it is not what we Americans definitively identify as "corn," the Native American "maize" (see footnote 16).

Now why was Layard thinking about food as he was digging into the ruins of Nineveh in 1845? Romantic, yes, but terrifying, times. They still are. The mound of Nineveh is found in the middle of Mosul in northern Iraq.

VII.
The Exchange of Peoples and Food

We call it a revolution in food, but it took much time and experiment to establish. And after its development, it took even more time to distribute across continents and within populations. Sometimes it even came back from where it started as a new, even more popular food. Two examples are instructive. One involves the spread of agriculture and animal husbandry into Western Europe. The second is what we describe as the great "Columbian Exchange" of animals, plants, and people between the New and Old Worlds.

The migration of agriculture into Western Europe from the east brought along a sophisticated culture of animal husbandry in addition to seed crops. And it was a culture that embraced the consumption of milk and milk derivatives. Milk contains a sugar, actually a two simple sugar combination called lactose, a "disaccharide." But enzymes that are present in infants necessary to breakdown that lactose in their mother's milk are lost after weaning. So a milk-based culture required the expansion of a gene continuing to produce through adulthood the enzyme capable of breaking down the lactose in milk. More than 60% of the world's human population is lactose intolerant. But in Europe, the agricultural revolution was accompanied by an increase of lactose tolerance in the population. And a great proportion of American citizens are descended from this European culture. We drink a lot of milk here.

The Columbian Exchange had an even greater impact on the world. It is defined by trade between New and Old Worlds. Where would the planet's population find the calories for growth without corn and the potato?[53] They are not native to the Old World. And what would American farms look like without pigs, cattle, and horses? They are not native to the New World. No cattle drives, no dairy farms, no "Old Kentucky" horse farms, and no "beasts of burden." Only the llama was available in the Americas and could only be used for small loads.

With the advent of trade through the Columbian Exchange, the availability of foodstuffs exploded. The world's population did too. The potato, whose many varieties still populate its land of origin in the Andes Mountains of South America, is a power pack of energy-rich carbohydrates.

Americans are well-versed in the Irish diaspora to America after the failure of that country's population-supporting potato crop. And potatoes were a life-saving crop during winter European wars. Armies, like locusts, consumed all foodstuffs within their reach. They were rarely interested, however, in digging further in the ground looking for potatoes! China is now the number one producer of "spuds."

The Columbian Exchange created a change in world culture and order. The American lands received new, larger animals capable of greater work, meat and dairy production. The Old Worlds, which included Asia and Africa, obtained high calorie crops of great yield. The populations in both areas surged. And what would Italian dinner dishes look like without the tomato? Another American import. And Asian dishes? No heat without these American chilies!

Yet there had been earlier trade between disparate human populations. The famous Silk Road, traversed or imagined by Marco Polo in 1271 A.D., had to cross a continuous, 3,000-mile long barrier of mountains from Manchuria to Afghanistan. Imagine the wonder of traveling from China along a 4,000-mile long trail, crossing the Gobi and

Taklamakan deserts around a chain of oasis, and standing before the great defile of the Dzungarian Gate.[54]

Into the fertile grasslands you would walk. Past the site of the 751 A.D. Battle of Talus, where the Muslim Abbasid Caliphate defeated the Chinese forces of the Tang Dynasty, eventually converting most of Central Asia to Islam. Then onto Samarkand, now in Uzbekistan, the "door" to trade on the Silk Road, and eventually the seat of power of Tamerlane and the Timurid Empire.

An even earlier route of trade between Asia, India and the Middle East (and hence Europe) involved coastal sailing from the southern tip of Saudi Arabia to India and its Asian trade connections. Eventually, Arab sailors developed navigational tools to cross the Indian Ocean directly. Yemen was more important than just a source for frankincense and myrrh then. It was a bustling center in this Asian trade. And the description of coffee as "Mocha" comes from the name of one of its ports. Both this trade route and that of the Silk Road would traverse Iraq, making Bagdad and the Abbasid Caliphate wealthy in trade.[55]

To a great extent, the Portuguese voyages of Vasco da Gama, which reached India in 1497, changed the flow of trade from the east to the west. And we cannot leave the global trade in foodstuffs without discussing the triangular trade in African slaves, Caribbean sugarcane and molasses, and New England rum.[56]

North America was also the recipient of foods carried in by migrants from Central and South America. A three-part plant complex called the "Three Sisters," spread into Eastern North America around 500 A.D. The assemblage was grown together in a single mound with each species complementing the other by providing favorable growing conditions. Eaten in combination, they provided nutrients that one alone lacked. We even honor the "Three Sisters" on the back of the Sacajawea Dollar.

As would happen later when New World corn reached the Old World, corn arriving in Eastern North America from the south, provided abundant calories resulting in new population growth here.[57] The great centers of the "Mississippian Period," Native American cities such as Cahokia, opposite St. Louis, sprouted up and flourished. It is not surprising that corn also supported the Great Mayan civilization of Central America, where their culture described a great "Maize Tree" from which humans formed. They placed a representation of this tree in their heavens. You can see it now, growing out of a crack in a turtle shell that you were foolishly taught was "Orion's Belt." The Mayans were perhaps the first true "People of the Corn."

Are plants such as corn facilitating human evolution and expansion? Or is the opposite true? Are we simply the tools for the expansion and evolution of the corn genome? It is a subject fun to consider.

We now have the investigative tools to trace back corn's genetic origins. Our corn, maize, was developed from a tall weedy grass in the highlands of Mexico. Teosinte is what it is called there now, and I have grown it in Kentucky. It requires a long growing season. But one cold October day I saw its short "silk" tassels form and, by first frost, a single row of triangular kernels had matured in a slender paper-like casing. Wow, the first "corncob!"

As the Three Sisters provided the extra calories for population growth, Native American societies abandoned the early gardening crops they had been selectively growing. The plants that had formed the earlier Eastern Agricultural Complex disappeared from both cuisine and the environment, although a Bolivian relative of goosefoot, Quinoa is now a nutritious and trendy health food.

There are many issues, facts and events that bear on our experience with food in history. Not all of them, obviously, can be included in this small discussion regardless of the author's intent. But I include, in various footnotes, a cornucopia of delicious entries.

VIII.
Grasses, Grains, Beef and CAFOs [58]

We used to classify them as "monocots," more modern forms of the great flowering (literally) of angiosperms. The first flowering plants were a revolution in reproductive success and they soon overwhelmed their predecessors as the dominant plants on land. Those earlier plants, the gymnosperms (including conifers such as pines, spruces, and hemlocks) still exist but generally inhabit more stressed environments. In cold and severe soils, like in the Canadian forests, angiosperm advantages are discounted and the gymnosperm adaptations become competitive. But it is in angiosperm monocot grasses, specifically the order *Poales*, from which the great proportion of the world is fed.

Energy from carbohydrates, carbohydrates from grasses. Small seeds embraced by big imaginations, feeding even bigger brains in evolving humans. Human minds, which independently across the globe, selected tiny seeds to become the basis of their civilizations.

The biggest components of that human feeding come in six grasses: corn (*Zea mays*), wheat (*Triticum sp.*), rice (*Oryza sativa*), barley (*Hordeum vulgare*), millet (a generic term for numerous species of small-seeded grasses such as pearl millet (*Pennisetum glaucum*), and rye (*Secale cereale*). So perhaps a Queen knew what her people really wanted when she allegedly said, "Let them eat cake." And Cleopatra was valuable to Caesar not only as a consort, but as a supplier of grain. History is all about Carbs!

And that doesn't even include the massive production of sugar from another grass, the sugarcane (*Saccharum officinarum*). And still to be told, the amazing saga of the potato. But those must be revealed in other stories. Too much for this meal.

The dog was the first animal domesticated by humans. The earliest evidence that we have for this domestication is from Iraq, about 12,000 years ago. Whether for hunting, protection, or just human companionship, all modern dog breeds track back to the wolf genome.

Sheep and goats were domesticated in Northern Iraq around 9,000 B.C. They were the first to be herded for food, including meat and milk. And, of course, sheep provided their hair as wool.

By 7,000 B.C., ox and cattle had become domesticated in Asia. They provided abundant meat, milk products, and, importantly, work. Power for transportation and the plow. The big developments of agriculture and animal husbandry fostered more permanent villages. Larger ones, too. Enter Hammurabi.

Cattle (including cows) and ox (in India, the bullock) are different species but are both considered "bovines," a subfamily of the mammal family, *Bovidae*. The Hindu religion generally restricts the slaughter and consumption of cows. The diversity of bovids explains how India, with a religious composition greater than 90% Hindu, can be a large exporter of what we call "beef" in the West.

So how did India become such a large exporter of beef? Water buffalo (*Bubalus bubalis*), another bovid. Water buffalo are in a different genus than our domesticated cattle (*Bos taurus*). Thus, the slaughter and export of water buffalo as "beef" is not prohibited on religious grounds in India. There are many genera of bovids, including the American bison (*Bison bison*), a species with a tragic history in America and one that, for Native Americans, was an important food source.[59]

In America, beef cattle are a major economic force, tying tightly into the production of corn and the dietary habits of the nation. Meat production in America is highly regulated through the laws and regulations of the United States Department of Agriculture. The outbreak of Creutzfeldt-Jakob Disease and "Mad Cow Disease" in England in the 1990s, gave us all concern. It led us to understand that a naked protein, the prion, was actually a form of life capable of reproducing in cattle populations. Prions are also present in deer populations where it is believed to be the cause of "chronic wasting disease." The United States Department of Agriculture (USDA) also investigates and orders recalls of American animal products contaminated with infectious disease vectors such as E. coli and Salmonella.

Concentrated animal feeding operations (CAFO) can include smaller animal feeding operations (AFO). They can involve cattle, swine, poultry, and seafood. Much of our animal foods can be concentrated in "Factory Farms." Industrialized agriculture. These operations speak much of our nature, needs, and habits. It speaks to costs, efficiencies, and environmental impact. And it speaks to our humanity. The Wondrous Journey of Food is not necessarily pretty.[60]

IX.
Religion and Agriculture

References to food are replete in the great writings of history. And these texts present culinary "laws" of what can and cannot be eaten, and what form is acceptable for consumption.[61]

Many religions developed from ancient cultures in the Middle East. Recent archaeological excavations in the area have uncovered evidence that an early prohibition on eating pork had developed in the highlands of Israel during the Iron Age. Pig bones are absent from fire pits and trash middens there from that time period.[62] And both the Bible and the Quran provide abundant prohibitions and culinary instructions. Much of their adherents' lives are measured by these strictures.

Cattle are a mainstay of western nutrition. The market for beef drove much of the human migration across the plains of America. But cattle are an ancient product of domestication, providing early advantages to agricultural communities of the Middle East and Central Asia. Beasts of Burden. A cow could provide work, milk, cream, butter, cheese, yogurt, meat, fuel (cow pies), and more cattle. The bull was revered in many ancient societies. For example, Gilgamesh kills the "Bull of Heaven." Moses returned from the Mount of Sinai to a "Golden Calf." The Roman cult of Mithras was based on the ritual slaughter of bulls. And there is always the "Bull Market" of the American Stock Exchange!

In the Hindu religion, cows are an important representation of Hindu beliefs concerning the presence of the Divine in all living creatures. The cow, with its afore-mentioned benefits to society, is the embodiment of such principle. As a result, many Indian states have restrictions on the killing of cows. The

Supreme Court of India upheld the validity of such laws in the Indian state of Gujarat.[63] As a result, cows in much of India are protected with sanctuaries provided for their care. These religious prohibitions, however, are not shared by all cultures in India. Their consumption of "beef" has sometimes resulted in political strife.

X.
Water, Salt and Fertilizer

Human beings are as much as 66% water by weight. We are creatures of water, evolved from development in a liquid medium. We are born in an amniotic sac of water-based fluids. Water is required for our cellular and bodily functions. Without it, we quickly die. Yet, submerged in water, we die even more quickly. Water, the basis of our lives, is a dangerous alien environment to us.

Water is a simple, yet beautifully acting molecule. In mass, it flows like a silken liquid lace. Because of its structure, molecules or minerals with an ionic charge dissolve within it and are evenly transported throughout its medium. Thus, the ability of water to move nutrients and wastes throughout the body. When these compounds splash up against the cellular wall, the mysterious Second Law of Thermodynamics takes over and transports water's cargo across the cell's semipermeable membrane. Diffusion and Osmosis. Magical processes.

Water floods throughout the human cell. It permeates the tiny and mighty mitochondria, the powerhouse of the cell, providing for respiratory release of energy in the consumption of sugar. From that combustion, Adenosine Triphosphate (ATP), the locomotive of life, quickly travels about the body, dispersing energy for growth and metabolism.

Yet, for all of the water in the world, most has too much salt and is, by the action of that same Second Law of Thermodynamics, dehydrating and toxic.

Migrate to strange lands and you will look for salt. Walk across the Cumberland Gap in Kentucky and you will look for salt. Find salt and you have found a home, for not only do you need salt for the proper function of your cells, you will need it to preserve food over winter. In an age of refrigeration, we forget that. But in lands with limited or no electricity, salt is still golden.

Yet the agricultural revolution was not enough. It built the launch pad and rocket ship for population growth, but still lacked enough fuel. Fuel to carry larger payloads of people. Unconstrained growth in feed stocks is necessary to support such an explosion in population growth. And the selective breeding of grains was not enough. There were other limiting factors.

Fresh water, yes. An immediate concern. But with language, coordination and ever-larger groups of workers, water could be brought to the crops through irrigation. And arid lands were often best suited to grow crops.

No, the real limiting factors are nitrogen and phosphorous. Nitrogen for the amine group in amino acids, and phosphorous for the backbone of genetic material. And pesticides were just as necessary as fertilizers for the modern "Green Revolution." We are not here in such great numbers by chance. We have "terraformed" this planet to support us.

XI.
Agricultural Production in the United States

The United States is comprised of diverse landforms with fertile and productive soils. The story of America's farm production is an exciting, romantic, and sometimes disturbing tale of the settling of this land. Think of the great western cattle drives, and the hunting to near extinction of the mighty bison. Think of the forced migration of Native Americans, the "amber waves of grain," and the "dust bowl." And think of sassafras, tobacco, corn, and cotton.

Sassafras was one of the reasons for the location of English colonies in America. A member of the *Lauracea* family of flowering plants, sassafras oils are aromatic. European plants lack these oils and, consistent with the doctrine of signatures, they were thought to be a cure for illness.[64] Yet the New World actually contained few new spices for the world's consumption.

Tobacco contains a powerful stimulant, which was quickly exported to the rest of the world. The regulation of tobacco production in the United States and the separate regulation of its labeling and consumption represent a uniquely American experience in culture and law.

Cotton, though not primarily a foodstuff, was the natural beneficiary of global trade systems. The importation, and eventual breeding, of human slaves in America to grow and harvest cotton led to a plantation-based system in America where slaves supported other agricultural production as well as animal husbandry. Much of America's agricultural history is reflected by and described in the Declaration of Independence and the United States Constitution. Yet arguments continued over the admittance of new territories as "free" or "slave" states and led to Civil War. Even today there seems to be a modern "politic" separating rural and urban populations. Is the location and production of foodstuffs still the defining character of our nation?

XII.
Kentucky's Experience

Kentucky is an exceptionally fertile and productive land. It is ancient land, barely touched by the glaciers. Its surface has been eroding and its soil forming since the great seas retreated more than 300 million years ago during the Pennsylvanian Period. Due to a great arching of rock caused by uplift of the Appalachian Mountains, Central Kentucky had eroded down to ancient Ordovician stone laid down more than 450 million years ago.

Kentucky has many unique resources. Some are relatively unknown. The native Paw Paw, *Asimina triloba*, is a delicious fruit of understory trees found in Kentucky's mesic valleys. There are currently programs to develop this fruit into a delicious agricultural commodity.

Unionids, or freshwater mussels, are abundant in Kentucky. In fact, the Green River of Kentucky was one of the centers of mussel evolution. More than 5,000 years ago, during the Archaic Period, Native Americans feasted on mussels in the Green River of Kentucky. The western portion of that river is replete with ancient trash piles of shells known as "shell middens."

Have you ever tried to eat a freshwater mussel in Kentucky? Good luck! It is hard to get the tough slimy meat from the shell, much less cut the muscle into strips. You should not even try since many of Kentucky's species are threatened or endangered.

And legal protection is in place to regulate mussel shell harvest. One hundred years ago, buttons were made from the shells of these unionids. And now tiny spherical cuttings of these shells are used as "starters" for the cultivated pearl industry. Although regulated, "midnight" mussel harvesting continues.[65]

Here is current data for crop production in Kentucky.[66]

XIII.
Environmental Law and Regulation

Alexander Hamilton! To him, in a great part, we owe our Federalist system of government. And to him also, we must attribute some of the angst over state rights when the federal government passes legislation of national application.[67] The individual states claim to know better. Let us do it. But when it comes to our food supply, the federal government has acted with many comprehensive national programs. Some deal with foods themselves, or additives thereto, and some with food safety. Some deal with farming and agricultural operations. Others deal with the impact of agriculture on the environment. And two of them deal with our most important food of all. Water!

There is a wonderful history to environmental law. Here in the United States, you can trace its development, examine its science, and judge the regulatory response to challenges facing our population. And when it comes to water, its progress is enlightening.

So much discovery and change occurred in America after World War II. It culminated in the exploration of new cultural freedoms and expressions in the 1960s. Americans believed that they could face any challenge and, with science, technology, and the law, conquer it. That newly embraced awareness led to the great efflorescence of federal laws addressing the environment that were passed in the 1970s. And leading the charge was President Richard Nixon and the "Rivers and Harbors Act of 1899" (Refuse Act).[68]

In 1970, in the face of an unpopular war and concern over contaminated waterways, President Nixon invoked the provisions of the "Refuse Act" by Executive Order requiring all wastewater discharges to navigable waters be permitted by the United States Corps of Engineers. Consistent with national interest in controlling water pollution, Congress passed the Federal Water Pollution Control Act Amendments (FWPCA, the Clean Water Act, 33 U.S.C. §§ 1251 et seq.), codifying much of President Nixon's action.[69] President Nixon vetoed the law as being too expensive. Congress overrode his veto, making this seminal statute law.

Water makes up about 66% of the human body and 75% of the human brain. In comparison, a tomato is 95% water and a chicken, 75%. The average person requires 2 1/2 quarts of water per day.

Of the Earth's surface, three-quarters is covered by water. Yet 97% of Earth's water is salty. Of the remaining fresh water, two-thirds is frozen in glaciers and ice caps. In the United States, 95% of the fresh water is under the ground.[70]

In 1974, Congress enacted the "Safe Drinking Water Act" (SDWA) (42 U.S.C. § 300F et seq.). The SDWA established maximum contaminant levels (MCLs) if "... the contaminant may have an adverse effect of the health of persons ..." (42 U.S.C. § 300G-1(b)1(A)). The SDWA standards apply to "public water systems." As the drinking water crisis in Flint, Michigan has demonstrated, the effects of drinking water on the populace can be acute and the cost to protect them, significant.

There are a number of federal environmentally-oriented statutes that deal with food and agricultural issues. They include many environmental laws, such as the Clean Water Act (FWPCA, above) and the Resource Conservation and Recovery Act (hazardous waste, 42 U.S.C. § 6901 et seq.). Yet many environmental laws exempt much of the agricultural activity. (for example see 40 CFR 261.4(b)(2)).

The Federal Insecticide, Fungicide, and Rodenticide Act (FIFRA, 7 U.S.C. §§ 136-136y) was originally passed in 1947 in response to the increase of pesticide use in agriculture spurred on by World War II. In 1962, Rachel Carson drew the attention of the world to the dangers of pesticides when she published her influential work, "Silent Spring." (Carson, 1962). Carson caught the attention of the nation. A quote from her book is noteworthy:

> *The history of life on earth has been a history of interaction between living things and their surroundings. To a large extent, the physical form and the habits of the earth's vegetation and its animal life have been molded by the environment. Considering the whole span of earthly time, the opposite effect, in which life actually modifies its surroundings, has been relatively slight. Only within the moment of time represented by the present century has one species, man, acquired significant power to alter the nature of his world. (Carson 1962, Chapter 2; but consider the impact of the Great Oxygenation Event 2.5 billions years ago).*

FIFRA was significantly amended in 1972. The amendments required the registration of new pesticides in the United States. There are now more than 1,600 categories of pesticides in use. Globally, more than five billion pounds of pesticides are used annually. Twenty-two percent of that total is used in the United States. EPA administers the program and takes into account risks as well as social and economic benefits. (Angelo, 2013).

The Federal Endangered Species Act was passed in 1973. This exceptional, forward-thinking legislation has withstood many attacks concerning its impact to private property rights. (ESA, 16 U.S.C. §§ 1531-1544; see also, *Tennessee Valley Authority v. Hill*, 437 U.S. 154 (1978)). A listing by the United States Fish and Wildlife Service (FWS) or the National Marine

Fisheries Service (NMFS) protects against the killing or taking of the protected organism and, in some cases, prevents the disturbance of its habitat. As agricultural production expands and the use of fertilizers and pesticides increase, the ESA becomes even more relevant. Will the spread of genetically modified organisms in agriculture further impact the natural plant and animal communities? Will the Endangered Species Act play an even more prominent role?

The Federal Food, Drug and Cosmetic Act was originally passed in 1938. It was significantly amended in the Food Quality Protection Act of 1996 and is administered by the Food and Drug Administration (FDA). (21 U.S.C. § 301 et seq.). The Food Safety Modernization Act of 2010 is also administered by the FDA and seeks to prevent and respond to contamination in our food supply.

In 1990, Congress passed the Organic Foods Production Act (OFPA), setting standards for "organic" products. (7 U.S.C. §§ 6501 et seq.) It is administered by the United States Department of Agriculture (USDA), as is the National Organic Program (NOP), a regulatory program overseen by the USDA's Agricultural Marketing Service. The NOP sets standards but does not regulate food safety or nutrition. (7 CFR Part 205).

Perhaps the most significant federal action involving American agriculture is what we call the "Farm Bill." This extensive economic and political legislation is promulgated approximately every five years. It is eagerly anticipated and is the subject of hearty controversy. The first Farm Bill was passed in 1938 as part of the New Deal. The most recent promulgation was entitled, "The Agricultural Act of 2014." (H.R. 2642; Public Law 113-79). The Farm Bill provides farm subsidies and agricultural price supports. Among other programs, this Farm Bill also supports the significant Supplemental Nutrition Assistance Program (SNAP). One source calculates the cost of the current Farm Bill at approximately 956 billion dollars.[71]

XIV.
Population

Food is about people. And people, at least in our species, are gregarious and aggregate at the family level. Put agriculture in the picture, and an even more settled lifestyle develops. Villages form. Soon, ambitions of infrastructure sprout. Irrigation, food storage, audacious temples, and warfare follow. For all of these developments - progress let us say - more people are needed. So many that one population can dominate its neighbors. So many that they get in each other's way. And then so many that they can't be fed. Or so Robert Malthus and Paul Ehrlich thought.[72]

The development of advanced genetic analysis has exploded the field of anthropological genetics and the study of our peoples. And the complete sequencing of the human genome has allowed comparison between different groups of human populations. The use of statistical analyses has allowed for the tracking of specific genetic traits within populations. Thus, scientists have recently identified the timing and direction of the migration of peoples across the planet.

The revolution of plant and animal domestication allowed for the first great expansion of the human population. The Columbian Exchange brought new, intensely caloric foods into global use. Just after World War II, the Green Revolution brought advanced plant breeds, chemically manufactured fertilizers and pesticides to enhance productivity. Malthus and Ehrlich were proven wrong, at least for now.[73] What next revolution will support the continuing rise of our population? And what will be the effect of Climate Change? Read on!

XV.
Climate Change

Much has been said of this testy subject. What fun! End of worlders, climate deniers, politicians ("I'm not a scientist"), hippies (yup, still out there), socialists, capitalists, scientists, and worst of all, statisticians. But climate change is real, always has been. Big Ice Ages, little Ice Ages, dust bowls, volcanoes, continental drift, and even meteor impacts. W o w , what a great place! So what's all the fuss about? What to do about it, I suppose. But for us, here in this discussion, we limit ourselves to its impact on food. And its impact will be as much about "where" food will be, as "what" food will be. Where is our food production going to migrate? What new areas will increase in agricultural fecundity and diversity?

Then there is that nagging problem of desertification.

And do you really think the seas are rising?[74]

There is science behind the analysis of climate change. Exciting science. Take, for instance, Milutin Milankovitch of Serbia. In 1920, he published his studies of multi-year ice age cycles and related them to changes in the Earth's eccentricity, axial tilt, and precession.[75] Good science, excellent observations. But there are other factors afoot impacting climate change. Does the flip of Earth's magnetic field have such impact? Although not regular, the shift occurs about every half million years. The North Pole migrates to the South Pole and vice versa during this phenomenon.

Sunspots are tremendously large and energetic events. The smallest visible sunspot is 50 times the size of Africa.[76] But we are 93 million miles away from the sun. What other factors might impact climate change on Earth? Well, here is a string cite of available factors - remember, some could cause warming of Earth, others cooling: methane (don't forget methane hydrates); soot (black carbon); hydrofluorocarbons (HFCs); volcanic action; sulfur dioxide; mountain building; solar flares; tropospheric ozone; nitrous oxide (N_2O); ocean conveyer systems; El Nino; La Nina; polar amplification; algal growth; cattle populations; aerosols; and, of course, the one receiving the most attention, carbon dioxide (CO_2).

In 1958, Charles Keeling began reporting CO_2 concentration in the atmosphere over Hawaii's Mauna Loa crater. In 1958, the global concentration was approximately 315 ppm. It is now over 400 ppm. CO_2 levels in the atmosphere track global temperatures fairly well, so the question arises; where is it coming from? Can such a small change in a minor constituent of the atmosphere- nitrogen and oxygen together make up about 99% - really trigger climate change? Do you care enough to change lifestyles and economies to address it? And was the "global warming hiatus" real?

The Intergovernmental Panel on Climate Change (IPCC) made some startling statements in its "Climate Change 2013: The Physical Basis" report.[77] And the prominent American Association for the Advancement of Science (AAAS) issued a review entitled "What We Know: The Reality, Risks and Response to Climate Change."[78] And isn't it getting warmer around here?[79]

The National Aeronautics and Space Administration (NASA) and the National Oceanic and Atmospheric Administration (NOAA) have determined that 2016 was the warmest year on record.[80]

XVI.
Genetically Modified Organisms

We live in a world with seemingly divergent interests. We wish to feed our massive populations, many of whom are undernourished, while also providing naturally healthy, wholesome, and "organic" food choices. We have already moved through one Green Revolution brought on by chemically generated fertilizer and pesticide production along with accelerated plant selection. And we have staved off the starvation earlier predicted.

But as our population on the planet continues to grow, will we need another revolution, a genetic one, to again increase agriculture production and efficiency? And can this be done in a sustainable, balanced way so as to protect the land and environment in which we live? Probably not, if our general record of past progress is any indicator. And now, with the release of genetically engineered "genies," and the creation of lamassu-like "chimeric" creatures, will we irreparably change nature? In ways we don't understand and can't control? But isn't that what we've always done?

It is these questions and more that make not for a "Brave New World," but for an "Unknown World." We tamper with the essence of life. Might we mistakenly produce Shelley's monster? So much yet to learn. Here follows some of the mechanisms by which we will accomplish these changes, along with a brief discussion of the nascent and confusing regulatory scheme that we have created to address them.

Genetic Engineering (GE) leads to "Genetically Modified Organisms (GMO)." You might think that such organisms would have an agreed upon regulatory definition. It is a reflection of the speed and intensity upon which this issue bears upon us, that such agreement is lacking.[81]

The genetic process whereby GMOs are produced is surprisingly diverse. And yet the structure of DNA, the "double helix" of James Watson, Francis Crick, and Rosalind Franklin, is amazingly simple. Chromosomes are composed of matching strands of deoxyribonucleic acid (DNA), each strand a simple stacking of 4 types of nucleotides, each nucleotide composed of a nitrogenous base, a 5-carbon sugar (deoxyribose) and a phosphate group. In DNA (ignoring for the moment RNA), there are only four nitrogenous bases: adenine (A), thymine (T), guanine (G), and cytosine (C). These bases have specific affinities: adenine to thymine; guanine to cytosine. So each nucleotide contains a simple variable code, with only four possible types of information (A, T, G, or C). Among other things, chromosomes code for proteins, primarily enzymes. Enzymes do most of the work in building and operating organisms. So, in order to provide the instructions necessary to build a life form, you must code for the building blocks of protein; amino acids. There are twenty of those amino acids needed for construction in humans. So we need a code arrangement with at least twenty different settings. Let's see, one nucleotide gives us just four; a group of two nucleotides gives us sixteen. Too small. A group of three nucleotides gives us sixty-four possible codes. More than we need, but it will do. That group of three is called a "codon," and a codon is the basic building code of genetics.

Change a nitrogenous base within a codon and you've likely changed the amino acid sequence in a protein. Add or delete a nitrogenous base and a similar change occurs. These are mutations, but can also be tools of genetic engineering. In fact, by using nucleases we can knock out whole series of codons, disabling deleterious proteins or changing their effect.

We can also rearrange codons on the chromosome. That is because chromosomes generally come in similar groupings. In our case, they come as "homologous pairs." Somewhere back in our evolution, the number of our chromosomes doubled, the nucleus retaining an extra copy of its genetic data. This development is common in life forms, and is an example of a mutation providing more material on which natural selection can work. So we humans have forty-six chromosomes in twenty-three pairs. We call that condition "diploid." Some species have three or four copies of their chromosome and are referred to as "triploid" or "tetraploid."

Now don't confuse the homologous-paired chromosomes with the double-stranded helix of DNA within a single chromosome itself. Nature uses only one of those strands for coding purposes. The second is a covering protection that "unzips" during "transcription" of the coded message into Messenger Ribonucleic Acid (mRNA).

Sometimes a break will occur in the limbs of adjacent chromosomes and the nucleus reattaches the loose pieces in the wrong order. Another mutation and another opportunity for variation. We also use this technique in genetic engineering.

In meiosis, the homologous chromosomes line up and are separated into the separate egg or sperm cells containing twenty-three chromosomes each (now haploid, referred to as gametes).

There is a history in science, as well as in law, and in both, excitement. The fact that so many discoveries in the field of genetic engineering were found in living organisms makes them even more intriguing.

By 1970, discoveries in molecular biology gained insight into the action of nucleic acids such as DNA (Deoxyribonucleic Acid) and RNA (Ribonucleic Acid).[82] Studies of the bacterium *Escherichia coli* and the movement of nucleic acids spurred our understanding of gene manipulation. The recognition that some cells contained independent loops of replicating sections of nucleic acids (called plasmids) that could transfer foreign genes into a host cell, was just one of such mechanisms discovered. Viruses, including bacteriophages, were also discovered to accomplish the same goal. *Agrobacterium tumefaciens* is a soil bacterium that causes plant tumors - called crown gall disease - in vegetable, fruit, and tree species. DNA in plasmids within the bacterium can be manipulated and transferred into the nucleus of another cell, e.g. a plant cell, to be joined with the cell's DNA where it will be added to the genome.

Cells naturally contain enzymes that maintain, copy, and repair DNA. Some of these have been identified, isolated, and copied so as to be usable in the laboratory. DNA Polymerase works to build and copy DNA from nucleotide building blocks. Restriction Enzymes (endonucleases) allow for the cutting of DNA at specific locations.[83] The Polymerase Chain Reaction (PCR) allows for a small quantity of a specific region of genetic material to be copied and multiplied.[84]

There is also now much interest in bypassing DNA altogether and inserting genetically engineered messenger RNA (mRNA) directly into a cell. Once there, the mRNA template can slide into the cell's ribosomes, which will assemble the protein products called for in the mRNA code. Think antibiotics, hormones. Lots of possibilities here. The trick is to keep the cell's immunity system from attacking.

The vehicles used for inserting foreign DNA into host cells are surprisingly varied. Plasmid DNA in bacteria is double-stranded, usually in circular form, and separate from the cell's chromosomal DNA. And plasmid DNA carries its own replication mechanism. These plasmids induce a number of reactions in their hosts, including beneficial traits such as nitrogen fixation as well as antibiotic resistance. Some are pathogenic, however, such as those inducing tumors in plants.

Bacteriophages are viruses that attack bacteria (prokaryotes). They are excellent vehicles for the transfer of genetic material. Yeasts are eukaryotes ("true cells"). The similarity of their cell structure and metabolism to higher forms of life, including human, make them useful for many genetic engineering procedures. As *Escherichia coli* has been thoroughly studied as a representative of prokaryotic cell metabolism, *Saccharomyces cerevisiae* (yeast, a unicellular fungus) has been studied for eukaryote analysis. This is the yeast common to both bread baking and wine fermentation.[85]

It is also possible to combine vectors such as plasmids and bacteriophages into hybrids called cosmids, phagemids, and fosmids. And finally, for our purposes, recombinant DNA can be directly transferred into host cells using such methods as "microprojectile bombardment," "electroporation," "UV laser irradiation," and "ultrasonication" through silicon carbide "whiskers," and by use of chemicals and heat shock.

An early application of genetic engineering to agriculture was the insertion of a gene with a valuable trait from one species into a receptor species that could benefit from that trait. So for example, look at the bacteria, *Bacillus thuringiensis* (BT), that lives in the soil. These bacteria have intracellular plasmids containing DNA molecules that operate independently of the regular chromosomal DNA. Within this isolated DNA are genes that code for proteins that are toxic to a broad spectrum of insects (lepidopterans, coleopterans, dipterans, hymenopterans) and some nematodes. Such toxins are natural pesticides within the bacterium that could provide protection from pests if reproduced within agricultural crops.

To engineer the transfer of such traits, the first step is to identify, isolate, and "clip out" the target bacterial gene through the use of restriction enzymes. These genetic strips can then be replicated in bacterial DNA or through the Polymerase Chain Reaction (PCR) to produce multiple copies. Then the gene can be physically forced through the membrane of a target species or carried inside as part of a bacterial plasmid or a virus. Once within the target species, the gene is incorporated into the host's genome. It will express the beneficial protein and provide protection to the entire plant. For research in animals, stem cells are often targeted to receive such gene transfer.

Genetic engineering activities are now widespread in industry and science. Corn is an example of a food plant that has been engineered to contain the BT gene, which provides pest protection during its growth. Another category of GMO plants includes soybeans, which have been genetically engineered to tolerate herbicides (HT) such as glyphosate, which interrupts amino acid metabolism in non-protected plants.[86] In addition, scientists are experimenting with manipulating plant and animal genomes to enhance desired traits.

A recent development in the field of genetic engineering involves "CRISPR." This acronym relates to primitive DNA sequences called "Clustered Regularly Interspersed Short Palindromic Repeats." When associated with RNA-guided Nucleases (CRISPR Associated Proteins, e.g. CAS9), pinpoint surgery can be performed on genomes, inserting, deleting, or rearranging gene sequences.[87]

Many food plants have been genetically engineered. Alfalfa (*Medicago sativa*), Argentine canola (*Brassica napus*), cotton (*Gossypium hirstum*), maize (corn, *Zea mays*), potato (*Solanum tuberosum*), soybean (*Glycine max*), sugar beet (*Beta vulgaris*), and wheat (*Triticum aestiuum*) have been engineered to resist the glyphosate herbicide. The "cry genes" of *Bacillus thuringiensis* (Bt toxin, Cry protein) target specific ranges of insect pests. These have been transferred into transgenic plants such as corn, cotton, and potatoes.[88] Crop plants can also be genetically engineered to express multiple types of resistance (stacked), providing protection against more than one herbicide and several different insect pests.[89]

Many other organisms have been genetically engineered. They include rice, jatropha (palm oil-like plant), yeast, cassava, papaya, banana, and tomato. A curious panoply of traits has been bred into plants, e.g. retarding fruit decay, antibiotic resistance, lignin production, drought tolerance, increased photosynthetic yield, male sterility, amylase changes, amino acid changes, flower color, reduced nicotine, and viral defense.

As a result, the cultivation of GMO crop plants is widespread. The leading producers of GMO crops are the United States, Brazil, and Argentina, followed by India and China. GMO crops are disfavored in countries of the European Market.

In 2015, the United States, Brazil, and Argentina grew 75% of the world's acreage in GMO crops. As much as 90% of the corn, soybeans, cotton, and canola crops in these countries has been genetically modified. In all, approximately 18 million farmers, in 28 different countries, grew 440 million acres of GMO crops in 2015.[90] In 2015, the United States grew maize, soybean, cotton, canola, sugar beet, alfalfa, papaya, squash, and potato crops that had been genetically engineered. Worldwide, soybean has 51% of the Biotech market, followed by corn (30%) and cotton (13%). Of genetically engineered plants, 53% are herbicide tolerant and 33% express "stacked" traits. Herbicide tolerance is the dominant engineered trait in soybean, corn, and cotton.[91]

The regulation of genetically modified organisms, GMOs, is relatively new, seemingly scattered, and perhaps insufficient to address the magnitude of change that faces us. Different schemes are applied in different countries.[92] The United States' experience in the regulation of GMOs is split between different agencies and different issues. To briefly describe that process, it is helpful to reflect on the Supreme Court case of *Diamond v. Chakrabarty*, 447 U.S. 303 (1980). There the court decided (5-4) that a bacterium containing genetically engineered plasmids transferred within was patentable under 35 U.S.C. § 101.[93] Value added I suppose!

In the United States, jurisdiction over GMO organisms is split between the United States Department of Agriculture (USDA), the Food and Drug Administration (FDA), and the Environmental Protection Agency (EPA). There are a number of policies and agreements seeking to regulate the activities of the agencies in this arena. The "Coordinated Framework for the Regulation of Biotechnology" was first promulgated in the Reagan Administration. In January 2017, the Obama Administration published two documents: 1) Update to the Coordinated Framework for the Regulation of Biotechnology;[94] and 2) National Strategy for Modernizing the Regulatory System for Biotechnology Products.[95] The White House also released a January 2017 statement relating to the National Strategy for Modernizing the Regulatory System.[96]

In January 2017, the Center for Veterinary Medicine of the FDA (part of the U.S. Department of Health and Human Services) issued draft guidance for the regulation of intentionally altered genomic DNA in animals (Guidance for Industry, #187).[97]

The Plant Protection Act (7 U.S.C. § 7701 et seq.) was enacted by Congress in 2000 to consolidate activities previously dealt with under the Federal Noxious Weed Act of 1974, the Plant Quarantine Act, and the Federal Plant Pest Act of 1957. Regulations can be found at 7 CFR 340 and are enforced by the United States Department of Agriculture

(USDA), Animal and Plant Health Inspection Service (APHIS). The Biotechnology Regulatory Services (BRS) deals with genetically engineered organisms under the Federal Coordinated Framework for the Regulation of Biotechnology referenced above.

EILERS

XVII.
Conclusion

I am still hungry. That is the general condition of all living things. Hungry. And there are lots of living things surrounding me in these woods along the streambed.[98] The stream itself is alive. For even in dry periods, the watercourse saves itself up in deep pools within dark bank undercuts or moist undersurfaces of limestone slabs.

Here, fish still dart around and crayfish scoot over algae-covered rocks. Underneath those rocks, a science fiction-like world exists, full of monstrous-looking creatures with shovel-like scoops to extend out and gather in prey. Some hide within tiny gravel-encrusted tunnels. Fish food. Dragonfly, damselfly, caddisfly, and mayfly larva. Truly an alien world to us, but vibrant and alive. No fresh fish food for us without it.

The warrior was an Algonquin speaker. His people hunted game, mostly smaller in those times, and gathered nuts and berries in their season. His people moved around in small groups, returning to certain rich bottomlands where they were experimenting with gardening, nascent agriculture. He, his wife, and child were back now. And they were hiding.

They were watching the strange man that had suddenly appeared in their fields and alongside the creek. Invaded, they were certain. But they couldn't figure him out. He did not seem to have a weapon, no flint-knapping kit. He made sparks easily, however, and that was troubling. What other tricks did he have hidden under that smooth animal skin he was wearing? They did not recognize such a pelt from any animal they had killed. This stranger, seemingly alone - and they had watched in hiding for most of two days - was full of unrecognized actions. The Native American's life was already full of the unknown. Change is scary. They held on tight to what their lives and culture had already proved.

The warrior rose quietly up and cocked his spear thrower directly to the back of his shoulder. Like me, he too was hungry.

SLAP! POP!

The dart entered the water in a small pool extending out from under the bank. The shaft quivered as the point hit and stuck in its mark. The warrior jumped into the edge of the pool, quickly sliding his hand down the shaft to the dart itself. He didn't want the dart to separate from the shaft and thus lose his prey.

There, got it! And the warrior pulled the writhing catfish from the water. With a wide grin, he turned to display it to me. It was an offering of food, a universal gesture, and it would prove to be a very tasty meal cooked over my fire.

I had jumped at the sound of the dart's impact, as had the other two, wife and boy. I hadn't heard the swing of the atlatl - the effect of the bannerstone I think - even though I had become sensitive to any sound around me.

Survival sounds. We ignore them now, but here many of those sounds were threats. The whoosh slap of the spear point scared me badly.

It took many primitive attempts at hand signals before we came together.

I will not ever forget that meeting that day and this place. Food became our communication, and through communication, my life. It became my place in the wondrous journey of food.

EILERS

XVIII.
Selected References and Resources

1.'Abdullah Yusuf 'Ali, *The Meaning of the Holy Qur'an*, Amana Publications, Beltsville, MA, 2001.

2.Allen, William H., "Thirst, Can Shipwrecked Men Survive if they Drink Seawater?" *Natural History*, December 1956.

3.Angelo, Mary Jane, and Jason J. Czarnezki, *Food, Agriculture, and Environmental Law*, Environmental Law Institute, Washington, DC, 2013.

4.Angelo, Mary Jane, "The Federal Insecticide, Fungicide and Rodenticide Act," *Food, Agriculture, and Environmental Law*, Supra, 2013.

5.Barlow, Connie, *The Ghosts of Evolution - Nonsensical Fruit, Missing Partners, and Other Ecological Anachronisms*, Basic Books, New York, NY, 2000.

6.Bell, Christopher L. et al., *Environmental Law Handbook, 22nd Edition*, Bernan Press, Lanham, MD, 2014.

7.Bishop, Bill, and Robert G. Cushing, *The Big Sort*, Houghton Mifflin Company, Boston and New York, 2008.

8.Braun, E. Lucy, *Deciduous Forests of Eastern North America*, Blakiston, Philadelphia, PA, 1950; reprinted by Hafner Publishing Company, New York, NY, 1972.

9.Bruggers, James, "Bracing for Climate Change" (a three-part series), *Courier Journal*, Louisville, KY, 2013.

10.Bruggers, James, "Bevin Wants Trump to 'Gut' the EPA," *Louisville Courier-Journal, Watchdog Earth*" November 10, 2016.

11.Bushdid, C., et al., "Humans Can Discriminate More than 1 Trillion Olfactory Stimuli," *Science*, Volll. 343, Issue 6177, pp. 1370-1372, March 21, 2014.

12.Carson, Rachel, *Silent Spring*, The Riverside Press, Cambridge, MA, 1962.

13.Chong, Key Ray, *Cannibalism in China*, Longwood Academics, Wakefield, IL, 1990.

14.Coe, Sophie D., *America's First Cuisines*, University of Texas Press, Austin, TX 1994.

15.Coe, Sophie D. and Michael D., *The True History of Chocolate*, Thames and Hudson Inc., New York, NY, 1996.

16.Committee on Scientific and Regulatory Issues Underlying Pesticide Use Patterns and Agricultural Innovation, *Regulating Pesticides in Food - The Delaney Paradox*, National Academy Press, Washington, DC, 1987.

17.Dalby, Andrew, *Dangerous Tastes - The Story of Spices*, University of California Press, Berkley and Los Angeles, CA, 2000.

18. Dalby, Andrew, *Food in the Ancient World from A to Z*, Routledge, London, England, and Routledge, New York, NY, 2003.

19. Darwin, Charles, *The Descent of Man, and Selection in Relation to Sex*, originally published by John Murray, London, England, 1871.

20. Department of Economic and Social Affairs, Population Division, *World Population Prospects, The 2015 Revision*, United Nations, New York, 2015.

21. Diamond, Jared, *Collapse - How Societies Choose to Fail or Succeed*, Penguin Group, Inc., New York, NY, 2005.

22. Diamond, Jared, *Guns, Germs, and Steel - The Fates of Human Societies*, W.W. Norton Company, New York, NY, 1997.

23. Ehrlich, Paul R., *The Population Bomb*, A Sierra Club - Ballantine Book, 1968, 1971.

24. Fernandez-Armesto, Felipe, *Near A Thousand Tables - A History of Food*, The Free Press, New York, NY, 2002.

25. Finkelstein, Israel, and Neil Asher Silberman, *The Bible Unearthed, Archaeology's New Vision of Ancient Israel and The Origin of Its Sacred Texts*, The Free Press, New York, NY, 2001.

26. Freidel, David, and Linda Schele et al., *Maya Cosmos - Three Thousand Years On the Shaman's Path*, Quill William Morrow, New York, NY, 1993.

27. Freud, Sigmund, *Totem and Taboo, Resemblances Between the Psychic Lives of Savages and Neurotics*, Vintage Books, New York, NY 1918, reprinted in 1946.

28. Gipe, Robert, *Unsuitable: The Fight to Save Black Mountain, 1998-1999*. See McSpirit, Infra, 2012.

29. Griffin, Nicholas W., and Philip P. Ahern et al., "Prior dietary practices and connections to a human gut microbial meta-community alter responses to diet interventions," *Cell Host Microbe*, Volume 21, Issue 1, pp. 84-96, 2017.

30. Heiser, Charles B. Jr., *Seed to Civilization - The Story of Food*, W.H. Freeman and Company, San Francisco, CA, 1981.

31. Howell, Catherine Herbert, *Flora Mirabilis - How Plants Have Shaped World Knowledge, Health, Wealth, and Beauty,* National Geographic, Washington, DC, 2009.

32. Imhoff, Daniel, Editor, *CAFO (Concentrated Animal Feeding Operation) - The Tragedy of Industrial Animal Factories*, Foundation for Deep Ecology, Sausalito, CA, 2010.

33. International Panel on Climate Change (IPCC), *Climate Change 2013, the Physical Science Basis*, 2013.

34. International Service for the Acquisition of Agri-Biotech Applications (ISAAA), *Brief 51: 20th Anniversary (1966 to*

2015) of the Global Commercialization of Biotech Crops; and *Biotech Crop Highlights in 2015*.

35. Karlawish, Jason, *Open Wound - The Tragic Obsession of Dr. William Beaumont*, The University of Michigan Press, 2011.

36. Kentucky Farm Bureau, *Kentucky Agriculture Facts Second Edition*, Kentucky Farm Bureau, Louisville, KY, 2015.

37. Kurlansky, Mark, *Salt - A World History*, Walker Publishing Company, Inc., New York, NY, 2002.

38. Lattimore, Owen, "Caravan Routes of Inner Asia," *The Geographic Journal*, Vol. LXXII, No. 6, 1928.

39. Laws, Bill, *Fifty Plants that Changed the Course of History*, Quid Publishing, England, 2010.

40. Layard, Sir Austen Henry, *Nineveh and Its Remains - A Narrative of An Expedition to Assyria During the Years 1845, 1846, 1847*, The Lyons Press, Guilford, Connecticut, 1882, 2001.

41. Lewington, Anna, *Plants for People*, The Natural History Museum, London, 1990.

42. Lomborg, Bjorn, *Cool It: The Skeptical Environmentalist's Guide to Global Warming*, Vintage Books, 2008.

43. Malthus, Thomas Robert, *An Essay On The Principle of Population As It Affects The Future Improvement of Society, With Remarks on the Speculations of Mr. Godwin, M. Condoret and Other Writers*, 1798.

44. Martin, Paul S., *Twilight of the Mammoths - Ice Age Extinctions and the Rewilding of America*, University of California Press, Berkeley and Los Angeles, CA, 2005.

45. McGee, Harold, *On Food and Cooking - The Science and Lore of the Kitchen*, Harper Collins, London, England, 1991.

46. McSpirit, Stephanie, et al. (Ed.), *Confronting Ecological Crisis in Appalachia and the South: University and Community Partnerships*, University Press of Kentucky, 2012.

47. Molina, Mario, *What We Know: The Reality, Risks and Response to Climate Change*, American Association for the Advancement of Science (AAAS), 2014.

48. Mueller, Tom, *Extra Virginity - The Sublime and Scandalous World of Olive Oil*, W.W. Norton Company, Inc., New York, NY, 2012.

49. Musselman, Lytton John, *Figs, Dates, Laurel, and Myrrh*, Timber Press, Inc., Portland, OR, 2007.

50. National Oceanic and Atmospheric Administration, *Global and Regional Sea Level Rise Scenarios for the United States*, United States Department of Commerce, January 2017.

51. Oxford University Press, *The Compact Edition of the Oxford English Dictionary*, Oxford University Press, 1971.

52. Palisas, Juli G. and Jon E. Keeley, "A Burning Story: The Role of Fire in the History of Life," *Bioscience* Vol. 59, No. 7, 2009.

53. Peabody, Carey E. and Christopher S. Heppner, 2016, "Groundwater Aquifer Depletion in Human Ecology," *Sustain, A Journal of Environmental and Sustainability Issues*, University of Louisville, Louisville, KY, Spring/Summer 2016.

54. Peattie, Donald Culross, *A Natural History of Trees of Eastern and Central North America*, Bonanza Books, New York, NY, 1948.

55. Plumer, Brad, "The $956 Billion Farm Bill, In One Graph," *Washington Post*, January 28, 2014.

56. Pollan, Michael, *The Botany of Desire - A Plant's-Eye View of the World*, Random House, Inc., New York, 2001.

57. Pollan, Michael, *Cooked - A Natural History of Transformation*, Penguin Group, New York, NY, 2013.

58. Pollan, Michael, *The Omnivore's Dilemma - A Natural History of Four Meals*, Penguin Books, New York, NY, 2006.

59. Ponte, Lowell, *The Cooling, Has the Next Ice Age Already Begun? Can We Survive It?*, Prentice-Hall, Inc., 1976.

60. Prothero, Donald R., *The Princeton Field Guide to Prehistoric Mammals*, Princeton University Press, Princeton, NJ, 2017.

61. Read, Piers Paul, *Alive*, Lippincott, Philadelphia, PA, 1974.

62. Robinson, John Mansley, *An Introduction to Early Greek Philosophy: The Chief Fragments and Ancient Testimony With Connecting Commentary*, Houghton Mifflin School, 1968.

63. Sandage, Tom, *An Edible History of Humanity*, Walker Publishing Company, Inc., New York, 2009.

64. Scallan E. Griffin PM, Angulo FJ, Tauxe RV, Hoekstra RM., "Foodborne Illness Acquired in the United States□Unspecified Agents," *Emerg. Infect. Dis.* Vol. 17, No. 1, in: http://wwwnc.cdc.gov/eid/article/17/1/p2-2201_article.htm, 2011.

65. Scallan, E. Hoekstra RM, Angulo FJ, Tauxe RV, Widdowson M-A, Roy SL, et al., 2011. "Foodborne Illness Acquired in the United States□Major Pathogens," *Emerg. Infec. Dis.* Vol. 17, No. 1, in http://wwwnc.cdc.gov/eid/article/17/1/p2-1101_article.htm.

66. Schlosser, Eric, *Fast Food Nation*, HarperCollins Publishers Inc., New York, NY, 2002.

67. Schneider, Stephen H., *Global Warming, Are We Entering the Greenhouse Century?*, Sierra Club Books, 1989.

68. Smith, Ruri Corley (trans), *The Last Japanese Soldier*, Tom Stacey LTD, London, England, 1972.

69. Snyder, Timothy, *The Black Earth, the Holocaust As History and Warning*, Tim Duggan Books, New York, 2015.

70. Sorge, Petra, "Where's the Beef in India Lately?" *Louisville Courier Journal*, December 31, 2016.

71. Steinhauer, Jennifer and Robert Pear, "Senate Leaders Agree on 2-Month Extension of Payroll Tax Cut," *New York Times*, December 16, 2011.

72. Tannahill, Reay, *Food in History*, Crown Publishers, Inc., New York, NY, 1988.

73. Tollenaar, Matthijs, et al., "The Contribution of Solar Brightening to the US Maize Yield Trend," *Nature Climate Change*, doi:10.1038/nclimatea3234, published online March 6, 2017.

74. United Nations, *World Population Prospects, the 2015 Revision*, Department of Economic and Social Affairs, Population Division, 2015.

75. Van Stockum, Jr., Ronald R., "Prehistoric Inhabitants," *The New History of Shelby County, Kentucky,* Shelby County Historical Society, Harmony House Publishers, Prospect, KY, 2003.

76. Van Stockum, Jr., Ronald R., "Population, Energy and Climate Change," *Sustain, A Journal of Environmental and Sustainability Issues*, University of Louisville, Louisville, KY, Fall/Winter 2015.

77. Van Stockum, Ronald R., Jr., "Human Ecology," *Sustain, A Journal of Environmental and Sustainability Issues*, University of Louisville, Louisville, KY, Spring/Summer 2016.

78. Van Stockum, Ronald R., Sr. *Remembrances of World Wars*, Chapters 31-34 "Guam;" Merrick Printing Co., Inc., Louisville, KY, 2013.

79. Viola, Herman J. and Carolyn Margolis, *Seeds of Change - Five Hundred Years Since Columbus*, Smithsonian Institution, 1991.

80. Wong, Edward, "Trump Has Called Climate Change A Chinese Hoax. Beijing Says It Is Anything But," *The New York Times*, November 18, 2016.

81. Wrangham, Richard, *Catching Fire*, Basic Books, New York, NY, 2009.

XIX.
Endnotes

[1] **E. Lucy Braun envisioned an ancient forest complex that survived the glacial ages in the Appalachian plateaus.** She believed that its greatest complexity was developed in the Cumberland Mountains of Eastern Kentucky and Western Virginia. Dr. Braun characterized this mountain assemblage as the "Mixed Mesophytic Forest." She described it as, "...the most complex and the oldest association of the deciduous forest formation. It occupies a central position in the deciduous forest as a whole, and from it or its ancestral progenitor, the mild Tertiary Forest, all other climaxes of the deciduous forest have arisen." (Braun, 1950).

On December 10, 1998, "Kentuckians for the Commonwealth" (KFTC) filed a "Lands Unsuitable for Mining" petition with the Kentucky Natural Resources and Environmental Protection Cabinet (NREPC). KFTC was represented by Kentucky Resources Council Environmental attorney, Tom FitzGerald. The petition sought to prevent mining on all land on Black Mountain above 3,000 feet in elevation, approximately 10,000 acres. Black Mountain was the type of location of Dr. Braun's Mixed Mesophytic Forest and supports the highest summit in Kentucky at 4,139 feet.

The petition was filed pursuant to Kentucky Revised Statute (KRS) 350.610, "Designation of Lands Unsuitable for Mining." For additional case law in Kentucky on this provision, see *Laurel Mountain Resources, LLC v. Commonwealth*, 360 S.W. 3d 791 (2012). The KFTC petition was resolved by agreement as referenced in an April 19, 1999 letter of agreement between the parties.

---

[2] **Few have looked up during summer and not recognized the star asterisms depicting a Teapot pouring tea on the tail of a Scorpion**. In Greek mythology, the Teapot is the constellation Sagittarius the Archer. It is in the Zodiac, an area of the sky through which the sun appears to move. The early Mesopotamians thought of the constellation as a centaur. I just see a teapot. No dispute over the next constellation, however. This grouping of stars looks like a scorpion and was seen as such by the Babylonians of the Middle East as well as the Mayans of Central America.

Much popular culture revolves around the astrological portent of one's birth sign. It is determined by the constellation that appeared behind the sun as it rose on the day of one's birth. These dates and signs were fixed by astrologers perhaps 1,000 years ago. Take a look at an inexpensive star application on a smart phone and set it back to your birthday. Run it for sunrise. You may be surprised by which astrological "House" you really belong. Your birth sign may have been changed by the wobble of the earth, precession. And that is a phenomenon that may also play into climate change!

---

[3] **A poem entitled "Steaming Tea."** It goes like this:

"To Greeks the claws that Mayans saw, that Romans tore and set afar, to hold the lunar scales for we, placed Scorpion on solar sea. Antares runs bright red like Mars, an archer aims a rising star, stinger poised yet barb restrained, it's only teapot pouring steam. Sagitta the Archer be, hunting Zuben Chimeli, the steam a Milky Way of stars within our world of spiral arms. A million more of galaxies expand our strings reality, yet only three from our world seen, grasp the Arab Genube. Xibalba Hades Hero Twins, the road beneath inflamed in sin, foolish men you dream to see, it's only teapot steaming tea." (Van Stockum, personal communication, 2004).

---

[4] **In fact, there are more miles of running water in Kentucky than any state other than Alaska**. At normal flow, there is more water from the Ohio River joining the Mississippi River at Cairo, Illinois than is in the Mississippi River itself at that point. Eastern North America is a verdant watery jungle!

Water is not so abundant everywhere else. California, which produces more than $50 billion in agricultural product, has only recently come out of a drought requiring statewide water rationing. And part of the genesis of the Syrian conflict in the Middle East was drought-driven human migration. As climate continues to change, those areas already stressed for fresh water are going to find obtaining sufficient supplies even more tenuous. And water is the stuff of life.

The Aral Sea, now lying between the Central Asian nations of Kazakhstan and Uzbekistan, was once the fourth largest inland sea. It is fed from the Amu Darya and the Syr Darya Rivers out of the mountains of Kyrgyzstan, the Tian Shan range. Cotton and rice irrigation has robbed the Aral Sea of more than 60% of its volume.

Nearby Lake Balkhash at the border of China and Kazakhstan faces a similar fate. It is the world's 15th largest lake and the one that travelers on the Silk Road reached through the Dzungarian Gate as they traveled west from the deserts of China into the grassland of the Semirechye and the seven rivers feeding the lake.

---

[5] **The 1970s was a decade when powerful federal environmental laws were promulgated.** In 1974, the United States passed the Safe Drinking Water Act (42 U.S.C. § 300(F) et seq.). Amended numerous times (including 1977, 1984, 1996, 2005, 2011, and 2015), this law seeks to protect the human population from nature, its own waste, and toxicity. It is expensive, however, and costs are generally borne by the taxpayers. As a result, legislation and enforcement is often reactive to crises in the water supply (e.g. Milwaukee, 1993, *Cryptosporidium*; Flint, Michigan, 2014, lead). For an example of this regulatory dilemma, see the findings of Congress in the 1996 Amendments (partially reproduced):

> "The Congress finds that –
>
> (1)safe drinking water is essential to the protection of public health;
>
> (2) because the requirements of the Safe Drinking Water Act (42 U.S.C. 300f et seq.) now exceed the financial and technical capacity of some public water systems, especially many small public water systems, the Federal Government needs to provide assistance to communities to help the communities meet Federal drinking water requirements; ..." (Pub. L. 104-182, Sec. 3, Aug. 6, 1996, 110 Stat. 1614-15).

---

[6] **Fresh water is not fresh, not necessarily clean, nor as clear as its cool silvery look imparts.** In the wild, animals contaminate the water with their use. And animals are often reservoirs for human pathogens. The parasitic flagellate protozoan *Giardia* is widespread in the fresh waters of the world. Microscopic and unseen, it survives in cysts after passing through human or animal feces. One who drinks from contaminated water may have unexplained weight loss,

massive flatulence, nausea, and abdominal cramps. No fun. That is why the Boy Scouts at Philmont, their high adventure camp in the Sangre de Cristo Mountains of New Mexico, disinfect the creek water there, careful to treat their bottle caps also. The treatment for elimination of the parasite from your body is no fun, either.

It is bad enough that fresh water can be contaminated in the wild. But it can be even more dangerous if it is contaminated with the water-borne diseases of humans. Untreated sewage, faulty septic tanks, "straight pipes," and ineffective wastewater treatment is widespread. *Salmonella*, *Entamoeba*, *Cryptosporidium*, *Giardia*, fecal coliform bacteria, cholera, dysentery, nematodes, flukes, and viruses (such as those causing hepatitis and polio) are only some of the risks. On that Silk Road into Central Asia, travelers drank boiled tea for water. (Lattimore, 1928).

Life is all around us. It is within us, also. There is a flourishing microbiota in our guts. These organisms consist primarily of anaerobic bacteria in a symbiotic relationship with our bodies, either mutualistic or commensal, along with fungi, protozoans, Archaeans, and viruses that prey on the bacteria. Our gut flora contributes vitamins (e.g. vitamin K and biotin) and can ferment organic compounds that we cannot break down. And what we eat and how much we eat changes the makeup of our intestinal passengers. Recent research has demonstrated that "... Americans consuming unrestricted diets maintained less diverse fecal microbiota than those of individuals adhering to a plant-rich diet with restricted caloric intake." (Griffin, et al., 2017).

---

[7] **There are three genera and four species of poisonous snakes in Kentucky:** 1) Copperhead (*Agkistrodon contortrix*);

2) Western Cottonmouth (*Agkistrodon piscivorus leucostoma*); 3) Timber Rattlesnake (*Crotalus horridus*); and 4) the Western Pigmy Rattlesnake (*Sistrurus miliarius streckeri*). You can identify all four of these snakes through characteristics that they share in the family *Viperidae*, subfamily *Crotalinde*. The Pit Vipers. That group also includes the deadly Bushmasters (*Lachesis sp.*) of Central and South America.

So here is how you can identify a native poisonous snake in Kentucky. First, they have vertical pupils, "cats' eyes" so to speak. Secondly, they have a pit between the nostril and eye. This is a heat-sensing organ for finding prey, the "pit" in Pit Vipers. Now, if you don't want to look close enough to see those two characteristics, you can always reach under and tickle the snake on its belly. You will probably get bitten, but in the off chance that the snake rolls over laughing, you will see that its belly plates (scutes) are entire below its anal plate. In a nonvenomous Kentucky snake, the scutes are entire above the anal plate and in two distinct rows below. So, here is a research assignment. If you are canoeing in Western Kentucky and a snake comes swimming by with its head held elevated out of the water, is it the poisonous Water Moccasin or the nonpoisonous, but incredibly ornery, Black Water Snake (*Natrix sipedon*)?

---

[8] **A flashlight. A wondrous object.** In this case, not a fancy halogen lamp with halogen gases and a hotter burning tungsten filament. Its emission peaks are in the red spectrum (approximately 800 nanometers). The flashlight in question is an LED (Light Emitting Diode) lamp. These diodes are electron-energized semiconductors, emitting photons of different energies. Original LED applications produced photons energized in the red spectrum. The Nobel Prize in

Physics for 2014 was awarded to three Japanese scientists (I. Akaskai, H. Amano, and S. Nakamura) for the invention of blue-emitting diodes. This enabled mixture with other diodes and a luminescent phosphor coating to produce white light. The efficiency of these new forms of artificial light, and the energy savings resulting therefrom, resulted in Congress phasing out the production of incandescent bulbs. See the Energy Independence and Security Act of 2007 (EISA), Pub. L. 110-140). The Act describes itself as:

> An act to move the United States toward greater energy independence and security, to increase the production of clean renewable fuels, to protect consumers, to increase the efficiency of products, buildings, and vehicles, to promote research on and deploy greenhouse gas capture and storage options, and to improve the energy performance of the federal government and for other purposes.

With a resurgence of opposition to the incandescent bulb phase-out after the 2010 Congressional elections, Congress passed the Consolidated Appropriations Act, 2012 (Pub. L. 112-74, H.R. 2055, 125 Stat. 785, December 23, 2011), which cut funding for the program. The *New York Times* reported that:

> The bill stipulates that no federal money can be used to 'implement or enforce' new energy efficiency standards for light bulbs. Republicans said the government had no business telling people what sort of light bulbs they could use.
>
> (Steinhauer, 2011).

---

[9] **Many states prohibit or regulate the killing of snakes without a permit** (see non-game code provisions in Georgia, Louisiana, Mississippi, Tennessee, and Virginia). In Arkansas,

a "depravation" permit is required to "take any game or fur-bearing animal damaging crops or personal property ..." Non-game wildlife, including snakes, can be taken without such a permit if they "... pose a reasonable threat to persons or property ..." (Ark. State Game and Fish Commission Code Book, Chap. 05 - General Hunting and Fishing Regulations, Section 05.10 - Depravation Permit Required). In addition, Kentucky prohibits the use of reptiles in religious services or gatherings; "Any person who displays, handles or uses any kind of reptile in connection with any religious service or gathering shall be fined not less than fifty dollars ($50), nor more than one hundred dollars ($100)." (KRS 437.060, Use of Reptiles in Religious Services (1942)).

---

[10] **Food is not just about taste.** We also know, of course, that it is about nutrition and, in great bulk, calories. The history of food is the story of how our ancestors obtained that food, lots of it, and efficiently. And it is that consideration of efficiency in which cooking plays a part. Your brain uses about 20% of your caloric intake, but for young children at rest it may be two or three times that amount. There are, after all, about 100 billion cells in the brain. So maybe more calories, more easily obtained, meant an evolution of bigger brains, human brains? One hypothesis surrounding this aspect of our evolution is called the "Cooking Hypothesis." It is described by Harvard Biologist Richard Wrangham in his 2009 book, "Catching Fire," and suggests that cooking prepares the energy stores in food so that less chewing is required and a quicker digestion provided. For example, if cooked food results in more quickly available absorption by the stomach and small intestine, then more nutrition will enter the bloodstream than would otherwise be absorbed by the flora in your colon.

Wrangham believes that cooking may increase digestibility by at least 10% and perhaps as high as 25-50%. As he points out, no aboriginal peoples exist without a cooking tradition, and that in the wild it would be difficult for a human to survive long on raw foods. Thus, it is possible that fire was tamed by our ancestors, and helped facilitate our brain size to begin its spectacular increase 1.8 million years ago. Cooking also renders food softer so less energy would be lost on chewing and tearing at it. That may explain the change in hominoid tooth structure and have provided longer life spans for the toothless elders.

Homo erectus, "Java Man" or "Peking Man" as the early skeletal assemblages were named, was the first truly human-like creature adapted to walk and live on the ground. The earliest fossil evidence for the creature is about 1.9 million years old. They appear to be the first of our kind to develop advanced stone, bifacial hand axes, and tools (Acheulean), and to have tamed fire in cooking hearths. They were the longest lasting species of the genus *Homo*, surviving almost ten times as long as our species has been extant. And they were still around 150,000 years ago!

Now, if cooking is what triggered larger brain size in the genus *Homo*, what is the earliest evidence for it in the archaeological record? The oldest data that we have is from two South African caves, Swartkrans and Wonderwerk. At those locations, there is sound evidence of our hominoid ancestors using hearth fires more than 1 million years ago. So the search for the origin of cooking fires continues to reach back toward the evolutionary origins of our bigger brains. And by the way, was human flesh on the menu for *Homo erectus*?

Charles Darwin said it best, maybe said it first, in his 1879 work, "The Descent of Man, and Selection In Regard to Sex."

For him, language and fire were the most important of human discoveries.

> He has discovered the art of making fire, by which hard and stringy roots can be rendered digestible and poisonous roots or herbs innocuous. This discovery of fire, probably the greatest ever made by man, excepting language, dates from before the dawn of history. (Darwin, 1871).

Cooking can also often render bitter and toxic foods palatable and can destroy disease vectors and pathogens. And that has been examined and legislated at both the federal and state levels.

The United States Food and Drug Administration (FDA) has since 1993 published a "Food Code" for the protection of the public.

> The Food Code is a model for safeguarding public health and ensuring food is unadulterated and honestly presented when offered to the customer. It represents FDA's best advice for a uniform system of provisions that address the safety and protection of food offered at retail and in food service. (FDA Food Code, 2013, P.i).

The legal status of the "Food Code" is described as follows:

> The model Food Code is neither federal law nor federal regulation and is not preemptive. Rather, it represents FDA's best advice for a uniform system of regulation to ensure that food at retail is safe and properly protected and presented. Although not federal requirements (until adopted by federal bodies for use with federal jurisdictions), the model Food Code provisions are designed to be consistent with federal food laws and regulations. ... (FDA Food Code, 2013, Preface iii).

The Food Code is updated by the FDA every four years and includes input from the Centers For Disease Control. The Food Code is more than 750 pages long and references estimate that "... food-borne diseases cause approximately 48 million illnesses, 128,000 hospitalizations, and 3,000 deaths in the United States each year. The occurrence of approximately 1,000 reported disease outbreaks (local, regional, and national) each year highlights the challenges of preventing these infections." (Scallan, 2011). Kentucky adopted and enforces the 2005 version of the Food Code as legal requirements in Kentucky through 902 KAR 45:005 entitled "The Kentucky Food Code."

Now on to the history and regulation of fire found in the next footnote!

---

[11] **Oxygen is life-giving stuff.** Breathe the air. The liquor of life. Heady. But it is also the agent of fire. Well, actually, it is fire itself. No oxygen, no fire. And we breathe oxygen in! And oxygen is a very mischievous atom. Eight protons, eight neutrons and eight electrons in two energy level clouds around the nucleus. Two electrons in the first level "S" orbital, two in the second level "S" orbital, and four in the second level "P" orbital. And therein lies the attraction of oxygen to other atoms, other matter. Oxygen needs two more electrons to complete that second "P" orbital. And it is aggressive in its desire to fill them. We call its action in doing so oxidation, rust, respiration, and fire!

Our ancient atmosphere did not contain oxygen in any appreciable quantity. All of the free oxygen had been tied up in the minerals and matter on this planet. Oxygen's appetite had been sated. Then early life evolved in our seas, evolved the ability to trap sunlight energy in its photosynthetic furnaces wherein carbon dioxide was forced into six-carbon storehouses of sunlight's fusion-generated energy; sugar glucose.

That early life form, hidden in the seas, generated a waste product. Most life does. In this case it was oxygen. Into the water it went, eating up (oxidizing) that with which it collided until an excess was able to escape to the atmosphere. We know when this occurred, about 2.4 billion years ago, because of the oxidation of iron sediments. We call these sediments, compressed into rock, the Banded Iron Formation (BIF) and the time of their creation, the Great Oxygenation Event. Go to Minnesota, you can see those rocks there.

About 21% of the atmosphere on Earth is now free oxygen. And that oxygen has risen to the stratosphere, where it intercepts deadly radiation, forming an invisible shield of ozone (O_3). This ozone layer protects the future evolution of complex life forms that sprung from those early photosynthetic cells. That's how we got here. Life terraforming Earth, preparing it for our arrival. Cool. Can we do that on Mars, too? Mars does not have an appreciable magnetic field like Earth. Even ozone might not be enough protection!

It turns out that there is a lot of oxygen around. I mean everywhere. It is the third-most abundant element in the universe. Eighty-six percent of the weight of Earth's oceans is oxygen and it is as much as half of the Earth's crust by weight. It is the most plentiful element on Earth.

And so now life has coated Earth's landscape with its organic paste and buried its carbon residue in coal, oil, and gas beds underneath. What difference does it make? Well, that residue will burn when ignited in the presence of oxygen and release that trapped carbon into the atmosphere. Will it warm the planet? That will be a discussion for another footnote. This footnote is ultimately about the controlled fire that we make through our enzymatic process of cellular respiration. There, oxygen and sugar are burned in the citric acid cycle, where it is converted to pyruvate and fuels the energy locomotive of our cells, adenosine triphosphate (ATP).

But what we colloquially call "fire" is the immediate goal of discussion herein. Fire, as Darwin noted in a previous footnote, is one of humanity's greatest discoveries. It cooks food, detoxifying some and making it easier to eat and digest. It provides heat on cold nights, scares off predators, clears fields, and is a weapon. But like the flooding that led to the English doctrine of Strict Liability, one wonders why law courts have not applied that doctrine to the most mischievous and wanton behavior of fire. Those characteristics were addressed in that seminal 1868 British Case, *Rylands v. Fletcher*, (LR3, HL 330 (1868) UKHL1):

> We think that the true rule of law is, that the person who, for his own purposes, brings on his land and collects and keeps there anything likely to do mischief if it escapes, must keep it in at his peril; and if he does not do so, is *prima facie* answerable for all the damage which is the natural consequence of its escape...

---

[12] **So what is the history of fire?** How did we tame it and how do we regulate it? As we have seen in the previous footnote, fire appeared on Earth when free oxygen became available, and a sufficient biomass had developed for which to burn. Plants began to spread on the land surfaces in the Silurian Period (approx. 440 mya). These conditions existed sufficiently in the Devonian Period (approx. 400 mya) for evidence of charcoal to have been found. As biomass on the land expanded later in the Paleozoic Period (approx. 345 mya), evidence of fire becomes more widespread. Fire became so common that many plants developed "serotiny," whereby seed germination occurred only after a fire. The concentration of

oxygen in the Carboniferous Period (approx. 350 mya, the age of our great Coal Forests) was as high as 31%, much higher than today's levels. A source of ignition was also necessary and could have been provided by lightning, volcanic activity, or even meteorites.

There are mythological tales describing the gift of fire to humanity. The Greek version involved Prometheus, the Titan who stole fire from Mt. Olympus by lighting a torch in the sun and smuggling out charcoal to the humans below. For this and other transgressions, he suffered mightily the wrath of Zeus. In revenge, Zeus chained a naked Prometheus to a cliff in the Caucus Mountains where by day vultures ate out his spleen. At night, his spleen regrew to extend his torment. More about vultures in another footnote. In order to save Prometheus, Pandora's Box was opened, releasing the Spites of Man, including old age, disease, jealousy, and hatred.

Undoubtedly, the first hominoids captured fire from wildfires and perpetuated it through hearth fires. Eventually, our ancestors learned to start friction fires with a wooden drill or by striking fine-grained stone (e.g. flint, chert, or obsidian) against a stone containing iron (such as the iron pyrite used by the Italian "Iceman" more than 5,000 years ago). Fire meant much to these early peoples. Cooking made food more nutritious and digestible. Fire-softened foods could be used to feed the elderly without teeth. Society formed around communal fires maintained for nighttime protection. So the early hominoids could finally abandon trees and sleep on the ground. And fields could be burned to maintain pastures, thus attracting game. Trees could be felled by girdling them with fire. And fire was a weapon that could be used against other humans. It continues in use today through the slash and burn agriculture of many undeveloped countries.

Hearths have been continuously maintained since those earliest of times. As braziers, they have long been essentials for travel:

> The poorer passengers seek shade or warmth, by burying themselves amongst the bales and other cargo, and sit patiently, almost in one position, until they reach their destination. They carry with them an earthen *mangal*, or chafing-dish, containing a charcoal fire, which serves to light their pipes, and to cook their coffee and food. ... (Layard, 1882).

In the Middle Ages, villagers were instructed by the ringing of evening bells to curb their nightly fires by covering them with what the French called "Couvre-feu" (cover fire). In 1068, William the Conqueror ordered the bells to ring the "Curfew" (modern English) at 8 PM and the population to bed.

In 1974, the United States Congress passed the Federal Fire Prevention and Control Act. 15 U.S.C. § 2201 et seq. Amended in 1990, the Act created the National Fire Prevention and Control Administration, the National Academy for Fire Prevention and Control, and the National Fire Data System. The 1974 law stated, in part:

> That Congress finds that - ... (3) fire is an undue burden affecting all Americans, and fire also constitutes a public health and safety problem of great dimensions. ... Fire-fighting is the nation's most hazardous profession. ... (11) The unacceptably high rates of death, injury, and property loss from fire can be reduced if the federal government establishes a coordinated program to support and reinforce the fire prevention and control activities of state and local governments. (15 U.S.C. § 2201, Public Law 93-498, § 2, Oct. 29, 1974, 88 Stat. 1535). (See also

40 CFR Part 257-3.7(a) entitled Federal Regulation of Open Burning of Solid Waste).

In Kentucky, the Energy and Environment Cabinet, Department for Natural Resources, regulates forest fire hazard seasons at KRS 149.400.

> (1) The periods commencing on February 15 and ending on April 30 and commencing on October 1 and ending on December 15 of each year are hereby declared to be and established as the fire hazard seasons. During the fire hazard seasons, even though the precautions required by KRS 149.375 shall have been taken, it shall be unlawful for any person to set fire to, or to procure another to set fire to, any flammable material capable of spreading fire, located in or within one hundred fifty feet (150') of any woodland or brushland, except between the hours of 6:00 p.m. and 6:00 a.m., prevailing local time, or when the ground is covered with snow. (KRS 149.400)

Open burning is also regulated by the Department for Environmental Protection, Division for Air Quality, and the Division of Waste Management. See 401 KAR 63:005. In addition, the Kentucky State Fire Marshall also regulates open burning. See KRS Chapter 227.220. Local ordinances also apply.

---

[13] **When you attempt to start a fire by striking flint against rock, the type of rock you strike will determine your success**. For it is not the flint that gives off the ignition spark. It is the tiny speck of iron from what is struck that becomes incandescent. With the advent of Iron Age "fire steels," metal-striking strips were carried against which flint or chert was struck. The flint, fire steel, and "char cloth" were often kept in a "tinderbox." A flintlock works on the same

principal, with the flint hammer snapping down against a steel "frizzen" plate. The resulting impact sparks into a small pan of gunpowder whose flame burns through a small hole, further igniting the powder charge within the barrel. "Half-cocked" is a colloquial phrase relating to its operation.

Matchsticks have their own interesting history, the first ones being dipped in sulfur. Other chemically soaked fibers produced the "slow" or "quick" match for guns, cannons, and fires.

Magnesium (atomic number 12, electron orbitals complete through the third level "S" orbital) is exceptionally flammable. It was the powder used for the photographic flash that you see in old westerns. In 1955, at the LeMan's Automobile Race in Monaco, more than 80 people died when a car with magnesium metal crashed and exploded into fire.

---

[14] **Flint is composed of silicon dioxide (quartz).** In sedimentary stone, it can be precipitated out of calcium carbonite (limestone). It is a mixture of ultra-fine grains of opal (colloidal silica) or chalcedony (cryptocrystalline). The silica may have come from the spicules of sponges, but flint formation is not well understood. Flint can be found as nodules in limestone beds, where it forms a characteristic patina when exposed to air. Since it lacks a regular crystalline structure, it fractures like glass to a sharp edge. It is hard enough to scratch glass. Obsidian, a volcanic form of quartz, lacks even the microcrystalline structure of flint. Its structure causes it to fracture even more like glass, forming edges so sharp that they can be used to slice objects thin enough to be used in electron microscopy. Obsidian is not native in Kentucky and was a prize import from the west for Native Americans making knives, scrappers, spear points, and

arrowheads. Flint used by Native Americans in Eastern North America can often be linked to quarries in specific geographic locations. Corydon, Indiana is one such location, and the Flint Ridge in Ohio, another.

Obsidian begins to form crystals and hydrate over time, so its structure will eventually change. What we commonly call glass is made by heating sand (again quartz, silicon dioxide), limestone, and soda ash at temperatures greater than 3,000 degrees F.

---

[15] **Flood plains are special places.** Fertile soil. Flat. Take the Nile, for example, before the Aswan Dam. Excellent building site. Wonderful water views. Except for the flooding. Some of our nation's most productive agricultural lands lie in the lower Mississippi River Valley, often flooding past the floodwalls. So here are competing forces; floods, and agriculture. And thus the need for laws and regulations to protect agricultural interests, especially when states and the federal government provide expensive disaster relief. More competing forces. This time, private property rights versus the police power of government bodies to regulate risks.

In 1965, a hurricane hit Louisiana, killing 75 people and causing more than $1 billion in damage. In 1968, the United States Congress passed the National Flood Insurance Act of 1968 (NFIP) (P.L. 90-448), 42 U.S.C. § 4001 et seq., which states in part:

> (a) Necessity and reasons for flood insurance program...
>
> The Congress finds that (1) From time to time, flood disasters have created personal hardships and economic distress, which have required unforeseen disaster relief measures and have placed an increasing burden on the nation's resources; ...

> (3) As a matter of national policy, a reasonable method of sharing the risk of flood losses is through a program of flood insurance which can compliment and encourage preventative and protective measures. (42 U.S.C. § 4001).

By 2010, the NFIP had sustained losses of almost $20 billion (this included costs associated with Hurricane Katrina in 2005). In 2010, there were 5.5 million properties insured by NFIP, with residential homes insured at $350,000, and non-residential properties at $1,000,000. In 2010, 50% of the homes insured were in Texas and Florida. And more than 25% of the residents of Florida and Louisiana participate in the program. Approximately 5.5 million properties are currently insured in the program.

Climate change with global warming is threatening coastal regions with increased flooding. In January of 2017, the National Oceanic and Atmospheric Administration released its report, entitled, "Global and Regional Sea Level Rise Scenarios for the United States," which includes the following partial statement in its Executive Summary:

> The projections and results presented in several peer-reviewed publications provide evidence to support a physically plausible GMSL [Global Mean Sea Level] rise in the range of 2.0 meters (m) to 2.7 m [approximately six feet, seven inchesto eight feet, ten inches], and recent results regarding Antarctic ice sheet instability indicate that such outcomes may be more likely than previously thought. To ensure consistency with these recent updates to the peer-reviewed scientific literature, we recommend a revised upper-bound scenario for a GMSL rise of 2.5 m [approximately eight feet, two inches] by the year 2100 ... (NOAA Technical Report Nos. CO-OPS 083; page vi).

The Federal Emergency Management Agency (FEMA) maintains maps of the nation's floodplains, which have a one-in-one-hundred chance of a flood every year. These 100-year floodplains are defined as Special Flood Hazard Areas (SFHA). FEMA will provide flood insurance to states that adopt regulations that are at least as stringent as FFIP requirements (see 44 CFR Chapter I, Subchapter B - Insurance and Hazard Mitigation, §§ 59-80):

> *Base flood* means the flood having a one percent chance of being equaled or exceeded in any given year. (44 CFR § 59.1)
>
> *Flood* or *Flooding* means:
>
> (a) A general and temporary condition of partial or complete inundation of normally dry land areas from:
>
> (1) The overflow of inland or tidal waters;
>
> (2) The unusual and rapid accumulation or runoff of surface waters from any source.
>
> (3) Mudslides (i.e., mudflows) ...
>
> (b)The collapse or subsidence of land along the shore of a lake or other body of water as a result of erosion or undermining caused by waves or currents of water ... or by an unanticipated force of nature, such as flash flood or an abnormal tidal surge ...
>
> (44 CFR § 59.1)

Kentucky's Division of Water requires flood plain development permits. KRS 151.250. The Division of Water Flood Mitigation Program has a 65-page "Quick Guide to Flood Plain Management in Kentucky" (2009; http://www.water.ky.gov). $206 million in insurance payments were made in Kentucky between 1978 and 2009. The highest numbers of flood claims are in Eastern Kentucky.

FEMA also produces a 585-page guide entitled, "National Flood Insurance Program (NFIP) Flood Plain Management

Requirements; A Study Guide and Desk Reference for Local Officials." (FEMA 480, February 2005).

---

[16] **This was the beginning of the great agricultural revolution** with the maturation of part-time gardening into deliberate selection of favorable plant characteristics, breeding, and cultivation. But it still took that silly visionary to convince those big game hunters that these little seed grasses held the key to their future survival, expansion, and power. It happened at different times in different places around the globe. Its effects came later to Eastern North America, arriving in force only around 500 A.D. with the development of temperate hardy corn from Central America. Those crops led to the development of the Great Mississippian civilizations like that at Cahokia on the Mississippi River across from St. Louis, or Wickliffe mounds in Western Kentucky near the junction of the Ohio and Mississippi Rivers.

Here, we will focus on Mesopotamia, the "Fertile Crescent" where, during the Neolithic Period 12,000 years ago, complex agriculture began. There, grasses led the way. Barley (*Hordeum vulgare*), and wheat (*Triticum spp.*), combined with various legumes such as lentil (*Lens culinarus*) and chickpea (*Cicer arietinum*, also known as the garbanzo bean).

The lands of Mesopotamia are fertile and need only irrigation to sprout great quantities of these foods. Large villages soon formed, generating significant disputes between its inhabitants. Enter Hammurabi. He wasn't the first "law giver," but he was an early one. He engraved his laws on a basaltic stele around 1780 B.C. in Babylonia, where he was King. There are 282 laws inscribed on the stele. The first one states, "1. Any person convicted of proffering charges against another person which he cannot substantiate shall be put to death." Number 282, the last law, states, "If a slave says to his master, 'you are not my master' and is proven guilty of this, his owner may cut off his ear." (Records of the Past, 1903, Vol. 2, Part 3). The Lillian Goldman Law Library at the Yale Law School has posted an online translation by L.W. King as part of their "Avalon Project, Documents in Law, History and Diplomacy." That translation mentions "corn" 55 times.

"Aye, there's the rub" (Shakespeare's Hamlet contemplating suicide). Corn was not available in the Old World until it was "discovered" in the New World by Columbus more than 3,000 years after Hammurabi lived. What gives? Examine the original Oxford English Dictionary (OED). My copy is a single volume with four pages reproduced on one page and sold with a magnifying glass. It appears that the original translators of Hammurabi's stele were British. Here is the OED definition of the word, "corn:"

Corn ...

I. *gen.* a grain, a seed ...

II. *gen.* Spec. the fruit of cereals ... (OED, 1971)

The same confusion (at least to Americans, where corn is corn) appears in Arabic, where the word for corn is the same word as that for grain, sonbol.

---

[17] **"Lebensraum," living space, as Adolf Hitler defined the term.** It was one of the prime objectives of Nazi Germany in World War II. Unlike the United States with western lands and its "manifest destiny," Germany only became a single nation in the late 1800s. By then, western nations had already established their expansive colonies. What African colonies Germany had in Africa, it lost in World War I. In the 1930s, Germany was unable to produce enough food for its population

and mobilize for war. According to one author (Snyder, 2015), this led to desire for the fertile agricultural lands of the east and the demonization of the Jews and Slavs located there. "For Germany ... the only possibility of a sound agrarian policy was the acquisition of land within Europe itself ... I need the Ukraine ... in order that no one is able to starve us again, like in the last war." (Citing Hitler in Snyder, pp. 15, 18, 2015). Although convinced of the power of science in the development of his war machines, Hitler did not believe, or want to believe, in the power of science to improve food production and yield. That "Green Revolution" was just around the corner and would have removed the need for calories as an argument for war.

---

18 **The human body cannot manufacture all of the organic compounds that it needs to function.** All too often we think of food only in the context of our caloric needs (energy) or the impact of various types of food on disease (e.g. saturated fats, and added sugars). But there are also essential organic compounds that we must ingest in our food and cannot manufacture in our body. Proteins are composed of amino acids. The essential amino acids needed from our diet are: 1) Arginine (youth); 2) Histidine; 3) Isoleucine; 4) Leucine; 5) Lysine; 6) Methionine; 7) Phenylalanine; 8) Threonine; 9) Tryptophan; and 10) Valine. Kwashiorkor is a disease of malnutrition caused by a deficiency of protein in the diet. There are also two essential fatty acids: 1) Linoleic; and 2) Alpha-Linolenic.

Essential vitamins obtained in our diet include: 1) Vitamin A (beta-carotin); 2) Vitamin B1 (thiamin) (prevents the disease Beriberi); 3) Vitamin B2 (riboflavin); 4) Vitamin B3 (niacin); 5) Vitamin B5 (pantothenic acid); 6) Vitamin B6 (pyridoxine);

7) Vitamin B7 (biotin); 8) Vitamin B9 (folic acid); 9) Vitamin B12 (cobalamin) 10) Vitamin C (as ascorbic acid) (prevents the disease scurvy); 11) Vitamin D (ergocalciferol, cholecalciferol) (prevents the disease rickets); 12) Vitamin E (tocopherol); 13) Vitamin K (naphthoquinoids). In addition, the following fifteen minerals are needed from the diet, either as electrolytes or as trace minerals: 1) Calcium (Ca) (atomic #20); 2) Chloride (Cl) (atomic #17); 3) Chromium (Cr) (atomic #24); 4) Cobolt (Co) (atomic #27); 5) Copper (Cu) (atomic #29); 6) Iodine (I) (atomic #53); 7) Iron (Fe) (atomic #26); 8) Magnesium (Mg) (atomic #12); 9) Manganese (Mn) (atomic #25); 10) Molybdenum (Mo) (atomic #42); 11) Phosphorus (P) (atomic #15); 12) Potassium (K) (atomic #19); 13) Selenium (Se) (atomic #34); 14) Sodium (Na) (atomic #11); and 15) Zinc (Zn) (atomic #30).

---

[19] **The Stone Age was hardly primitive.** The progression from chipped stone tools to the sharp slicing edges of obsidian was magnificent. The development of uniface edge stones (Oldowan) through the beauty and utility of bifacial hand axes (Acheulean) and composite blade tools (Microlithic) led to the Neolithic Period (New Stone Age). Small-grained rocks like flint and obsidian were worked into use as blades, spears, darts, and arrowheads. The evolution from hand axes, full groove axes, the larger and later three-quarters grooved axes, through to the smooth celt, gave early North Americans command over their environment. Along with fire, they felled trees, cleared fields, made housing material and boats. You should be so lucky to have such tools and weapons if you were found alone and abandoned.

---

[20] **Agriculture is the topic of this paper.** And for each of its elements, there is a history. Better law through history. We are now people of the corn, the potato and of the tomato. And all three of these crops are from Central and South America. Line out a graph of population growth against time. Is it just a coincidence that population on the planet exploded after the spread of these plants across the globe?

Corn was finally bred to succeed in temperate climates and arrived in Eastern North America as "flint" corn around 500 A.D. The impact of this temperate hardy corn was immediate and fed the furnace of population growth that archaeologists call the Mississippian Period. This was the time of the great civilization at Cahokia in Missouri, across the Mississippi from St. Louis, where 10,000-20,000 people lived. Cahokia is listed as a World Heritage Site by the United Nation's Educational Scientific and Cultural Organization (UNESCO). It is also protected by state law. (See the Archaeological and Paleontological Resources Protection Act, 20 Illinois Compiled Statutes 3435, 1990, and 17 IAC 4190; and the Human Skeletal Remains Protection Act, 20 Illinois Compiled Statutes 3440, 1989, 17 IAC 4170 (31); and 20 Illinois Compiled Statutes 3420, 199, IAC 4180). In 1985, the Illinois Historic Preservation Agency (IHPA) was created.

But I digress. It is the agricultural practice in Eastern North America prior to the arrival of temperate hardy corn that is the topic of this footnote. That nascent agriculture is called the Eastern Agricultural Complex. It was a gardening practice adjunct to hunter-gatherer societies. The plants cultivated were, for the most part, seeded, with the exception of squash (*Cucurbita_pepo*), which was grown for gourds and seeds. The seed plants were harvested for two different nutritional requirements. Sunflower (*Helianthus_annuus*) and sumpweed (marsh elder, *Iva annua*) were grown for oily edible seeds.

Starchy seeds were collected from maygrass (*Phalaris caroliniana*), erect knotweed (*Polygonum erectum*), little barley (*Hordeum pusillum*), and goosefoot or lamb's quarters (*Chenopodium berlandieri*). These gardens were supplanted by the arrival of The Three Sisters, temperate hardy corn, beans, and squash from Central America. Look on the back of the American Sacagawea dollar. There they are, The Three Sisters.

Beans are an important nutritional companion in corn diets. Corn is deficient in the essential vitamin, niacin (Vitamin B3, $C_6H_5NO_2$, precursor of NAD and NADP). Without niacin, the symptoms of the disease pellagra develop. Beans are relatively high in niacin. The "nixtamalization" of corn is a process developed in Mesoamerica. It makes the bound niacin in corn more available. It involves soaking corn in lime (calcium hydroxide) and ash (potassium hydroxide). Hominy grits are nixtamalized corn, as is the "masa" flour sold in grocery stores. Nixtamalization allows corn to form a dough.

Corn is also deficient in the essential amino acids, lysine and threonine, which are readily available in beans. Amino acid deficiency in primarily corn diets can lead to the disease Kwashiorkor. Corn provides good sources of the two essential amino acids, methionine and tryptophan, which are deficient in beans. The two crops complement each other.

And what of those garden plants of the Eastern Agricultural Complex that were being grown in Kentucky before the arrival of The Three Sisters? Wild birds and humans still eat sunflower seeds, but the original gourd species has vanished. The other seed species have also vanished or can only be found as roadside weeds. The cousin of one is now the trendy health food, quinoa (*Chenopodium quinoa*), a native of the Andes.

---

[21] ***Helianthus tuberosus*, the Jerusalem artichoke.** A sunflower. Wild, native to North America and full of the polysaccharide inulin, a polymer of the simple sugar, fructose. Delicious. Like a nutty potato. Sweeter if you let the inulin breakdown to fructose in storage. If not, it will just pass through. We lack the digestive enzyme to break down inulin, hence it acts as a fiber. But bacteria in our large intestine can digest it. Hence the gas. Low glycemic index, too. The name is probably of Italian origin, "girasole," meaning sunflower.

---

[22] **As noted in footnote 4, there is a lot of water in Kentucky.** That is a precious statistic in a country rife with massive population concentrations living on what are essentially deserts. Certainly many populations exceed the carrying capacity of their land. Take Denver, Colorado and Beijing, China, both which capture water from distant watersheds and divert or pipe it into their cities. Los Angeles survives on water from the eastern mountains and the Colorado River, 242 miles distant (think Jack Nicholson in the movie Chinatown). And, of course, the diversion of Hetchy Hetchy Valley water from the Sierra, Nevada Mountains to San Francisco broke John Muir's heart. Fly over the Hetchy Hetchy and look down at the dams. It was the gorge just north of Yosemite.

The Kentucky Water Resources Research Institute (KWRRI) was established by the Federal Water Resources Research Act, 42 U.S.C. § 10301 et seq. The KWRRI is housed at the University of Kentucky. Its website lists the following projects: Floyd Forks watershed, stakeholder involvement, water distribution system, security projects,

Kentucky Stormwater, Virtual Observatory Ecological Informatics System (VOEIS), Kentucky Total Maximum Daily Load (TMDL), Paducah stakeholder engagement, the Bayou Creek metals study, the Kentucky Nutrient Model, and the Pride Water Quality Project.

The Kentucky Division of Water administers water resources legislation in the Commonwealth of Kentucky through KRS Chapter 151. KRS Chapter 151.110 is entitled, "Water Resources Policy - Duties of Cabinet." KRS 151.140 is instructive in its regulation and exceptions thereto. It is entitled, "Withdrawal of Water from Public Waters, Permit Required -- Exceptions."

> No person, business, industry, city, county, water district, or other political subdivision shall have the right to withdraw, divert, or transfer public water from a stream, lake, ground water source or other body of water, unless such person, business, industry, city, county, water district or other political subdivision has been granted a permit by the cabinet for such withdrawal, diversion, or transfer of water. Provided, however, no permit shall be required for and nothing herein shall interfere with the use of water for agricultural and domestic purposes including irrigation; and no permit shall be required if the amount of water withdrawn, diverted or transferred is less than the amount established by regulation and no permit shall be required for water used in the production of steam generating plants of companies whose retail rates are regulated by the Kentucky Public Service Commission or for which plants a certificate of environmental compatibility from such commission is required by law, or water injected underground in conjunction with operations for the production of oil or gas. (KRS Chapter 151.140).

[23] **Corn. So much to talk about.** We know where it came from, both geographically and genetically. Its progenitor was native to the Highlands of Mexico, where its descendants still grow it today as a weedy grass called Teosinte. I have grown Teosinte in Kentucky. It grows tall and looks like a cornstalk. I planted it in early spring but never saw a cob forming. I figured that it needed a longer growing season when one week before frost, I saw short, scraggly looking silks seeming to float out of joints. Little faith I. When I pulled back the base of the corn leaf clasping tightly to the stem, I found it! The earliest form of a corncob. A single row of six or eight little triangular kernels tucked neatly together in a tight line. That was a pretty cool moment. I was holding in my hand the history of my people. People of the corn.

Corn is a grass, as are the other significant food grains. As a student, it was important to learn that grasses were monocots, more modern evolutionary descendants of the dicots. As with so much of our plant and animal taxonomy, the science of genetics is discerning new and surprising evolutionary paths of life's development and progress. But there is something else fun and interesting about corn. It is a C4 plant, not a C5 one. That is a big deal. Really.

Photosynthetic plants create glucose, a six-carbon simple sugar, from carbon dioxide, water, and sunlight energy. There are three pathways that different types of plants utilize to feed carbon dioxide into the photosynthetic Calvin Cycle through use of the rubisco enzyme. Rubisco is considered the most abundant protein in the world. The resulting three-carbon intermediate (glyceraldehyde 3-phosphate) can then be converted to sugar and starch. The three types of pathways that can feed carbon dioxide into the photosynthetic system are C4, C5, and CAM (crassulacean acid metabolism). The differences involve the cellular location of the photosynthetic steps, the build-up of oxygen, and the loss of water through the plant leaf's stomata. Most plants are C5 plants. CAM is used by desert plants and is efficient in its use of water. C4 plants, including corn and sugarcane, are more efficient in water use relative to sugar production, but more importantly, can photosynthesize at a faster rate in high sunlight, high oxygen-generating conditions. About 1% of the world's plants use the C4 method of metabolism. It is this difference in plant photosynthetic mechanisms that makes for an interesting consequence of global warming and increased levels of carbon dioxide in the atmosphere.

---

[24] **Freud and food. Totem and taboo.** Sigmund Freud developed intricate and surprising theories of psychoanalysis. One of them involved the eating, or prohibition thereof, of a "totem animal" in primitive societies. Of course, what is old is new so to speak, and Freud was interested in how these basic or primitive instincts persisted in modern culture. He described the "totem" as a sacred object that must not be killed or eaten. Yet, at important times, according to Freud, a "totem meal" performed an important function within a clan.

> The clan is celebrating the ceremonial occasion by the cruel slaughter of its totem animal and is devouring it raw - blood, flesh and bones. The clansmen are there, dressed in the likeness of the totem and imitating it in sound and movement, as though they are seeking to stress their identity with it. ...
>
> Psychoanalysis has revealed that the totem animal is in reality a substitute for the father; and this tallies with the contradictory fact that, though the killing of the animal is as a rule forbidden, yet its killing is a festive occasion - with the fact that it is killed and yet mourned.The ambivalent

> emotional attitude, which to this day characterizes the father-complex in our children and which often persists into adult-life, seems to extend to the totem animal in its capacity as a substitute for the father. (Freud 1918, pp. 193, 202).

Food, therefore, at least in the "totem meal" sense, may serve an important cultural and psychological function in our society beyond nutrition.

---

[25] **We have evidence of fire pits at the Old Bear Site in Shelby County, Kentucky.** These remains are part of the Newton Phase of the Late Woodland Period (ca. 510 A.D.). The remains found there might give you an idea of what's available to eat should you find yourself in the predicament described at the beginning of this book.

> The many fractured animal bones that were found probably represent marrow extraction. Remains of twenty-five animal species were found, including seven white-tailed deer, three black bear, five raccoons, two wild turkeys, and one individual each of gray squirrel, woodchuck, skunk, wolf or dog, beaver, soft-shelled turtle, and terrapin. Plant remains in the pits included squash, honey locust, butternut, shagbark hickory, walnut, and acorns. The presence of the broken bones of one adult human near the site is intriguing. (Van Stockum, Jr., 2003)

---

[26] **The wild turkey is native to North America.** Not so the chicken (*Gallus gallus domesticus*). The chicken is native to Asia and reached us through India.

Let us "talk turkey" and not be "chicken" to address the truth. Let us make a wish on a turkey's "wishbone." Turkey is an American bird and it's getting bigger all the time. Come on, you recognize their "carnucles," "snoops," and "wattles." How else would you separate the common American wild turkey (*Meleagris gallopavo*) from the other American species, the ocellated turkey (*Meleagris ocellata*)? True, the ocellated turkey is only found today around the Yucatan Peninsula. That is why it is represented in ancient Mayan drawings. And did you know that, other than the wild turkey, there was only one other bird domesticated in the New World; the Muscovy duck (*Cairina moschata*). That one never made it big on the world food table, but it is still popular with tropical farmers. I will let you figure out the "Muscovy" name for yourself.

By the way, do you know how much bigger the turkey on your table is today compared with 1960? Well, then it was 16.8 pounds in average weight. In 2014, it had risen to 30.4 pounds! A breeder tom is even bigger, between 50 and 70 pounds. For comparison, in 1960 the average American man weighed 166.3 pounds. In 2015, 195.5 pounds. Women in America now average 166.2 pounds, an 18.5% increase from 1960. And both sexes are about one inch taller.

---

[27] **Humans describe the use of energy in different ways.**

" In food, we use kilocalories (1,000 small calories) but call them Calories (KCALS). Check out any food label. Here is what it means. There are 4.2 joules (international unit of energy) in each small calorie.

A small calorie is the energy needed to raise the temperature of water 1.8° F (1° C). In one sugar molecule there are about 680 KCALs of which respiration retrieves

37-50% and traps it in the phosphate bond of adenosine triphosphate (ATP).

That is approximately 288 KCALs contained in 36 ATPs per sugar molecule (glucose). That is about 8 KCALs per ATP molecule in the last phosphate bond. About 8.8 ounces of your body weight at any one time is in ATP and represents about the energy in one "AA" battery. But every day you will convert and recycle your body weight in ATP. One alkaline based "AA" battery contains about 9360 joules (or 2.2 KCAL equivalents)."

(Van Stockum, Jr., "Human Ecology," footnote 27, 2016).

---

[28] **We commonly refer to light as those frequencies interpreted by our eyes and brains as sight.** Yet visible light is just a small slice of the wide spectrum of electromagnetic radiation of excited photons, all moving at the "speed of light." These photons carry varying amounts of energy, thus accounting for the different attributes of different frequencies within the full spectrum of electromagnetic radiation. This spectrum, and energy transfer within it, explains in part the "greenhouse effect" on Venus and the current global warming phenomenon on Earth. More on that later, but in the classic model, this radiation is described as continuous waves of various wavelength, frequency, and amplitude. Our world is permeated by these wavelengths and they are passing through your brain and body in floods of frequencies right now. If your brain was so configured, you could watch television, listen to the radio, and sense changes in far away objects. Are they dangerous? Reality is full of field perturbations, so who really knows? Concern, reasonable or not, has been expressed over the use of handheld cell phones.

Here is a summary of the electromagnetic spectrum from longer wavelength, lower energy radiation to the shorter wavelength, higher energy (these are dangerous ones) frequencies. At the low end of the spectrum, wavelengths can be as long as one-quarter the distance between the Earth and moon. The distance shortens to subatomic distances on the other end of the spectrum. The low-end, long wavelength spectrum moves to shorter wavelength power and telephone bands, radio waves, microwaves and then through infrared to visible light. Then, as the wavelength shortens and energy increases, you move through the ultra violet, x-rays, and gamma rays. Much of the sun's energy penetrates through the Earth's atmosphere on visible wavelengths, only to be absorbed by the planet and reradiated as heat in the longer wavelength infrared band. Therein lies much of the story of global warming. Greenhouse gases can more easily "trap" the re-radiation as heat.

---

[29] **Snakeroot, *Eupatorium ruguson* (now *Agertina altissima*), is a native Eastern North American perennial plant.** It contains tremetone, a chemical that is toxic to horses and concentrates in the milk of cattle. It is this "milk sickness" that may have killed Abraham Lincoln's mother in Indiana in 1818. The disease causes weakness, stiffness, severe constipation, bad breath, and then coma. It is listed in the "FDA Poisonous Plant Database," which states "white snakeroot can be eliminated from pastures by persistently pulling the plant each year." (FDA #: F04968). Lucy Braun (see footnote #1) has a snakeroot named for her (*Agertina lucide-brauniae*) that is restricted to the Cumberland Plateau.

---

[30] **Guam is the largest (210 sq. miles) of the Mariana Islands, a chain of isolated sea mounts in the Western Pacific Ocean.** It was the scene of fierce battles in World War II between Japanese forces and United States Marines. My father was there and has written about his harrowing experience (Van Stockum, Sr., 2013). The Japanese invaded Guam (then a U.S. territory) in December 1941. The United States Marines recaptured it in July 1944. Japanese Lance-Corporal Shoichi Yokoi eluded detection and capture, and for 28 years lived in isolation and in a cave dug out in the ground. In January 1972, two fishermen discovered him and he was finally brought out of the wild. What and how he ate during those long years is the subject of this footnote.

Yokoi often dreamed of food, stewed meat, broiled eel and snake. But there were no feasts in his dreams. Maybe such dreams warn the sleeper of famine and propel action. At one point, Yokoi stole a chicken and ate it raw. First he cut off the head and drank the warm blood. Then he threw away the intestines, eating the raw flesh and gnawing at the bones. The dreams worked.

Initially no fire was made for fear of being discovered. Any meat he could store was wrapped up in banana leaves. It soon became maggot-infested but he dug out the maggots with a stick and chewed through the rotting meat.

Yokoi had to be careful for he suspected that two other refugees had died from food poisoning, perhaps from eating the fruit of the palm-like Fredeico Nut Cycad Tree (*Cycas micronesia*). The seed, although nutritious, is rife with cyanide and must be split and soaked many times to leach out the poison. It has been linked to Lytico Bodig disease. Still, it was until recently a traditional food of the island. There are also poisonous toads on the island. Yokoi learned to avoid these. So much for the kindness of nature.

When he did use a fire, Yokoi at first used a flashlight lens to focus the sunlight into an ignition source. The lens broke, and he resorted to spinning a stick on a board to create fire. Then he devised a length of twined palm fiber rope, which smoldered when lit. This "slow match" was also soaked in oil for a lamp. From the fumes of the lamp, he discovered that he could smoke and preserve meat.

Breadfruit (*Artocarpus alilis*) was available from May to October. The large tubers of wild yams (*Dioscorea spinosa*) could be dug in December and January. The coconut tree (*Cocos nucifera*) bore fruit year-round.

Yokoi was able to capture wild pigs periodically with a noose, and even stoned pigs from a tree. He had greater success digging and camouflaging five-foot deep pits. He caught fish in a bamboo basket with coconut meat as bait, which he placed in local streams. He would harvest his baskets at midnight so as not to be seen. Yokoi also fashioned a coconut-baited mousetrap and would fry snails, lizards, and eels. One of his delicacies was the plump liver of a field mouse. He never used salt, fearful of going to the beach for saltwater and being discovered. (Smith, 1972).

After 28 of years hiding on Guam, Shoichi Yokoi returned to Japan in 1972. He died in Japan in 1997, aged 82. He was never introduced to Emperor Hirohito.

---

[31] **Turkey vultures (*Cathartes aura*, which means "cleansing breeze") and black vultures (*Coragyps atratus*) are both native to Kentucky.** I have a multi-generational turkey vulture family living in the upper floor of my old dairy barn. My children were raised with them. Hear the strange "ssss" of a juvenile and you will think about having been transported to another planet. But friends they are to you and

me. The Cherokee called turkey vultures "peace eagles" because they did not kill their prey. Black vultures do feed off small livestock afterbirths and still-borns, but they lack the talon strength to lift and fly away with food like hawks. My family is always pleased to see a vulture atop our broken-down silo or old dead behemoth oak tree with wings outstretched to the wind (the horaltic pose). They look like priests preaching to their flock, but I suspect that they are desiccating any parasites that found their way into their wing feathers.

They are like that. Fussy about cleanliness. That is why they have no feathers on their heads. They don't want to pick up anything extra when they are rooting around in the abdominal cavity of a dead animal. Did you know that they vomit on their legs? Same reason. They will sometimes urinate on their legs to kill and clean any hitchhikers they may have picked up down there. And vultures will vomit corrosive stomach acid in defense. Groovy.

It is against Kentucky law to kill a vulture without a permit. That is because both species of vultures in Kentucky are protected by the Federal Migratory Bird Treaty Act of 1918. (16 U.S.C. § 703-712). The law incorporates portions of treaties with Canada, Mexico, Japan, and Russia. The Supreme Court reviewed the implications to state law in upholding the treaty in *Missouri v. Holland*, (1920), 252 U.S. 416. The list of migratory birds protected by the Act can be found at 50 F.R. 1013. A description of listing history can be found in the March 1, 2010 *Federal Register*, 75 F.R. 9281, entitled "General Provisions; Revised List of Migratory Birds." The list now encompasses 1,007 species of birds.

---

32 **The Roman Empire had an affinity for meat and fish.** Unfortunately, much of that food tended to rot in the hot

Mediterranean sun. Age, shall we say. So even though it might still be nutritious, a lingering odor was not especially attractive. Enter the Roman's favorite sauce, liquamen or garum. Same stuff probably. Better ingredients in one versus the other. Have you ever smelled Thai Fish Sauce? That is all you will smell. That is probably the smell of liquamen. Yuk. But as a measured ingredient, an unbelievably tasty addition. The Romans put it on everything.

Here is how they made it. Scrape out the pieces of fish from your net, and cover them with salt. Then put the mixture out in the sun for several months. You will then have an awful, rotten liquid mess. Strain out the bits of tissue left and pour it on your favorite not-so-fresh food. Viola! (As the later Gauls might say).

There is another great food additive of Roman times, but it is a bit more of a mystery. Actually, it is a complete mystery. Silphium. Another one for you to research. From North Africa and then gone. You search for it.

---

[33] ***Brassica oleracea* is known as wild cabbage and is native to Europe.** Humans have cultivated and selected traits for consumption using all parts of the plant. The following foods all derive from this plant: Brussels sprouts, cabbage, broccoli, cauliflower, collard greens, kale, kohlrabi and rutabaga.

---

[34] **Drupes, nuts, fruits, vegetables, and tubers. Herbs.** Colloquial terms with a botanical basis. But the law sometimes applies a more practical definition. See *Nix v. Hedden*, (1893), 149 U.S. 304. In this interesting case, the Supreme Court was

required to determine whether a tomato was a vegetable or a fruit under the Tariff Act of March 3, 1883. The lower court found that, for the purpose of the Act, (which excluded fruits), the tomato was a vegetable.

Botanically speaking, tomatoes are the fruit of a vine, just as are cucumbers, squashes, beans, and peas. But in the common language of the people, whether sellers or consumers of provisions, all these are vegetables which are grown in kitchen gardens, and which, whether eaten cooked or raw, are, like potatoes, carrots, parsnips, turnips, beets, cauliflower, cabbage, celery, and lettuce, usually served at dinner in, with, or after the soup, fish, or meats which constitute the principle part of the repast, and not, like fruits generally, as dessert. (Nix, Supra).

---

35 **The Monarch Butterfly (*Danaus plexippus*) feeds on milkweeds (*Asclepius spp.*).** The butterfly concentrates cardenolides, heart-attacking steroids. So you see, there is a reason to reject bitter-tasting foods. Butterfly-eating birds have figured this out.

---

36 **The Federal Water Pollution Control Act Amendments of 1972 (FWPCA) were a direct outgrowth of President Richard Nixon's expansion of the permit authority available under the Rivers and Harbors Act of 1899 (the Refuse Act, 33 U.S.C. § 407).** Although President Nixon vetoed the Act as too expensive, his veto was overridden and the law continues to protect the nation's waterways as the Clean Water Act (33 U.S.C. §1251(a) et seq.). Here are some of the goals and policies set out in the Act:

(1) it is the national goal that the discharge of pollutants into the navigable waters be eliminated by 1985; ...

(3) it is the national policy that the discharge of toxic pollutants in toxic amounts be prohibited; ...

(7) it is the national policy that programs for the control of nonpoint sources of pollution be developed and implemented in an expeditious manner so as to enable the goals of this chapter to be met through the control of both point and nonpoint sources of pollution. (33 U.S.C. § 1251(a))

That's where fathead minnows (*Pimephales promelas*) and water fleas (*Ceriodaphnia dubia*) come into play. Bio-monitoring. Looking for toxicity in whole effluent. See 40 CFR 136.3. It is a tribute to our people that the nation continues to pursue the lofty goals set out in the 1972 FWPCA. Maybe fathead minnows and water fleas are found in the stream described in this story.

---

[37] **The Pleistocene Geologic Epoch of the Quaternary Period is the period of glaciers in North America.** We date its beginning at approximately 2.5 million years ago, concluding with the retreat of the glaciers about 12,000 years ago. We call the next Epoch that followed the Holocene. We like to call it the Age of Humans.

In the Americas during the Pleistocene, there were giant creatures extant on the land. Mammoths, mastodons, gomophotheres, ground sloths, toxodons, and glyptodonts. These creatures are gone now. Whether due to human hunting, climate change, or both, they are extinct. But they leave behind memories of their majestic presence. Trees with special fruit and seeds upon which they fed. Those trees have continued to reproduce, even though their dispersal agents have vanished. They are a shimmering mirage of what once was, but is no more. Few animals can eat the large fruits, large seeds, and sometimes bitter contents. In Kentucky, they include the Kentucky Coffee Tree (*Gymnocladus dioicus*), Honey Locust (*Gleditsia triacanthus*), Paw Paw (*Asimina triloba*), Persimmon (*Diospyros virginiana*), and Osage Orange (*Maclura pomifera*). Those big Paw Paw seeds were meant to travel through the gut of large creatures, dispersing far and wide in their already fertilized bed of dung. Now the creatures that feast on the Paw Paw fruits steal the pulp and leave the seed. A sad fall for once so mighty trees and their praetorian guards!

---

[38] **The Vagus nerve, Latin for "wanderer," runs from the cerebellum down to your abdomen.** It is your stomach talking to you. "Gut instinct," "gut reaction," "listening to your gut," "visceral feelings." Who hasn't felt "butterflies" in their stomach and become "sick to their stomach" with anxiety and fear? Hunger involves the vagus nerve, hypothalamus, the hormone ghrelin, neuropeptide Y, leptin, and pro-opiomelanocortin. But in the absence of a meal, I find that it is the shrinking of the stomach that fires up the vagus nerve the most. Have you ever sneezed when you were hungry?

---

[39] **Groundnut is *Apios americana*, a perennial vine also known as hopniss.** Native Americans relied on hopniss for nutrition, especially in the winter.

There are six native species of wild grape in North America: summer grape (*Vitis aestvalis*); sand grape (*Vitis rupestris*); frost grape (*Vitis riparia*); fox grape (*Vitis lubrusca*); mustang grape (*Vitis mustanensis*); and the muscadine (*Vitis rotundifolia*). Most wine is made from cultivars of the European species, *Vitis vinifera*, which leaves only one other species to be accounted for, that being *Vitis amurensis* in Asia.

In Kentucky, wines are made from native *Vitis lubrusca* grape; American-French hybrids between *V. vinifera*, *V. ruprestris*, and other European cultivars. In Kentucky, small family wineries are licensed through KRS 243.155 and defined at KRS 241.010 (5).

Moonseed is the poisonous climbing vine, *Menispermum canadense*. All of the plant contains toxicity, including the seeds. They are crescent shaped, unlike the round seeds of the wild grape. Also, grapes have tendrils, which moonseeds lack. Be careful what you eat in the wild!

---

[40] **1. Native dandelions (*Taraxacum spp.*);** 2. The common blue violet (*Viola sororia*); and 3. Wild leeks (ramps) (*Allium tricoccum*).

---

[41] **This is the atlatl, the one Aztec weapon feared by Cortez.** Its efficiency was recorded by Conquistador Bernal Diaz Del Castillo in 1520 A.D. He wrote of the atlatl, the Aztec word for the weapon in their native Nahuatl language, in his book, *A True History of the Conquests of New Spain* (published posthumously in 1632). The atlatl is a spear thrower capable of repetitive fire and sending an obsidian-tipped dart through chest armor. That's why Cortez feared it. Arrows bounced off their metal cuirass, but atlatl darts penetrated into the flesh below. Eskimo hunters still use them and they are featured in the opening scenes of the movie,

"Apocalypto," by Mel Gibson. The atlatl is composed of a two-foot hand-held launching stick (the spear thrower), a long arrow-like shaft, and a short detachable dart with a sharp, flaked spear point hafted at its end. The use of this spear thrower developed after the demise of the megafauna at the close of the glacial ages 12,000 years ago, and before the use of the bow and arrow in Kentucky around 500 A.D. What most artifact collectors call "arrowheads" are actually atlatl points or the blades from earlier, true spears.

---

[42] **The throwing stick, or spear thrower portion of the atlatl has a stone weight tied to its middle.** The advantage of weight to the physics of the throw is debated, but in the Southeast, it developed further, changing shape. Here a winged form, the "banner stone," developed. These flat, wide stones had holes hand-drilled through the middle of the stone's axis through which the spear thrower slipped. These bannerstones appear to add a noise-cancelling effect when swung. "Atlatl Bob" Perkins made an authentic reproduction of a southeastern atlatl for me. Its sound in operation is much quieter than those with the more linear banner stones used out west. So, not only is it a rapid-fire, high-caliber firing mechanism, it is a stealth weapon!

---

[43] **Carbohydrates (simple sugars); lipids (fatty acids); proteins (amino acids); nucleic acids (nucleotides).**

---

[44] **Major Elements in the human body by mass and relative percentage:** 1) Oxygen - 65%; 2) Carbon - 18%; 3)

Hydrogen - 10%; 4) Nitrogen - 3%; 5) Calcium - 1.5%; 6) Phosphorus - 1.0%. Water (H_2O) ranges around two-thirds of the weight of the human body and that accounts for the high mass of oxygen. Life is based on carbon-based compounds. But why don't we have skeletons made of silicon? Some sponges do, and that may be one of the sources of silicon found in flint nodules of sedimentary rock. You know, the ones early Native Americans "knapped" into tools and weapons.

---

45 **And carbohydrates are everywhere.** They come as monosaccharides of simple sugars such as glucose, galactose, and fructose; disaccharides such as sucrose (glucose and fructose; table or grape sugar), lactose (glucose and galactose; milk sugar) and maltose (glucose and glucose; malt sugar); and polysaccharides of many simple sugars joined together such as starch (e.g. potato) and glycogen ("animal starch," a storage reservoir in animal liver and muscle). The exhaustion of glycogen is often referred to as "hitting the wall" by runners. Plants also produce a polysaccharide we call cellulose, which humans can't digest. Microflora in the guts of horses and cattle can do so and they enable those animals to forage and eat hay. Insects also produce a polysaccharide, chiton (insect shell), for their exoskeleton. We can't digest that either, but it is rather crunchy, and insects can be quite nutritious!

---

46 **Except certain areas in Africa.**

---

47 **Democritus was a Greek philosopher who lived around 400 B.C. in Abdera Thrace.** He expounded on the

theory of atomism. He also related his theory to explain the sense of taste. For Democritus, taste was related to the basic shapes of atoms freed up by chewing; sweet was due to round atoms; sour, jagged atoms; pungent, fine, round, angular atoms; salty, large, twisted, two equal sided atoms; bitter, small, curved, smooth ones; and oily, small and round atoms. (Robinson, 1968).

---

48 **Mixing ingredients that stimulate the five senses of taste leads to the diverse foods, and cultures, found in our societies.** But how many truly different tastes can you make? Not so limited with the sense of smell. The human nostril contains four hundred different smell detectors. One laboratory performed tests that indicated that humans can detect at least one trillion odors! (Bushdid, 2014). That is what makes fine dining truly a garden experience.

But here is a curious condition. Perhaps 4.5% of the population has a variety of "Synesthesia," where senses are merged. Seeing colors when tasting or smelling, for example. Creativity runs with this condition and many who have it are artists.

---

49 **See this quote from a 1958 article in *Natural History*, a journal founded by New York's American Museum of Natural History.**

"At the beginning of dehydration there is thirst and discomfort.

Succeeding symptoms, in order, are lassitude, loss of appetite, sleepiness, rise in body temperature and, at about 5 per cent dehydration, nausea. At from 6 to 10 per cent dehydration, the victim will experience dizziness, headache, tingling in the limbs, dry mouth, difficulty in speaking, and inability to walk.

At more than 10 per cent dehydration, delirium is common, and the senses fail. Dehydration of 25 per cent is probably fatal at any temperature. At air temperatures above 90 degrees F., 15 per cent dehydration is the theoretical fatal limit. ...

... as dehydration increases, the victim's will to resist the desire to drink sea water weakens until finally he succumbs to the temptation and death is caused by the ingestion of sea water. ...

... There is immediate slaking, followed quite soon by an exacerbation of the thirst, which will require still more copious draughts. The victim then becomes silent and apathetic, with a peculiar fixed and glassy expression in the eyes. The condition of the lips, mouth, and tongue worsens, and a peculiarly offensive odour [sic] has been described in the breath. Within an hour or two, delirium sets in, quiet at first but later violent and unrestrained; consciousness is gradually lost; the color of the face changes and froth appears at the corners of the lips.

Death may take place quietly: more often it is a noisy termination, and not infrequently the victim goes over the side in his delirium and is lost. (Allen, 1956)."

---

[50] **Umami, delicious, yummy, savory.** The French chef Auguste Escoffier may have been cooking to it in the late 1800s. But Kikunae Ikeda, a Japanese chemist, discerned its essence in "dashi," a Japanese savory soup of seaweed. He discovered that it was based on the presence of glutamic acid. Ikeda was the one to name it in a chemical journal more than 100 years ago. It is the "L" form of glutamate that is most powerful. Anchovies are packed full of it. Why else would

you eat one? Okay, maybe you like salt, too. Fermented soy sauce gets you there, also. Ikeda also pioneered the use of monosodium glutamate, MSG, as a flavor enhancement. Or you can just eat a ripe tomato. Mouth-watering. A subtle pleasant after-taste will remain. Umami. Have another!

There are receptors on your tongue for L-glutamate. But why? Your body needs to ingest amino acids that it cannot manufacture itself. Free glutamate acid, whether from cooked meat, aged cheese, fermented soy, or ripe tomato, tells the body that digestible protein is available. Amino Acids. Your tongue encourages you to continue to feast. And Umami is enhanced by the presence of certain nucleotides. The building blocks of our genetic system, its transcription and translation, and energy transfer. The sensation of taste is not for your pleasure. It is for your survival. Enjoy!

---

[51] **Cannibalism.** A shocking word. Dangerous, too, as we will see. We have recorded it in nature. Black widow spiders and preying mantises. But it appears that cannibalism is a widespread strategy in the animal kingdom. See for example the lace-weaver spider (*Amaurobius ferox*), or tiger sharks. Better for you to research that for yourself. And debate has surrounded Peking Man and Java Man (*Homo erectus*) and whether associated bones are evidence of cannibalism. It appears that cannibalism may have been widespread at different times (Chong, 1990). It occurred in Asia, Europe and the Americas. See the famous discussion of the Donner Party in Western North America, or the stranded Uruguayan rugby team. (Read, 1974).

Some primitive tribes of New Guinea are also believed to have engaged in the practice. That practice involved funerary cannibalism where relatives ate the rotted meat of the corpse.

In addition to the meat, the brains of the deceased were mixed with ferns and roasted in bamboo tubes. Women were sometimes primary consumers because it was believed that they could better control the powerful spirits of the dead. In the 1950s, it was discovered that many in the population were dying of a mysterious ailment. They called it "Kuru," laughing death. In 1976, a researcher won a Nobel Prize for demonstrating that the disease was transmitted in brain tissue. Only later was that agent of disease discovered to be the prion, a "proteinaceous infectious particle." Think "Mad Cow Disease" in England, "Wasting Disease" in American deer populations, and Creutzfeldt-Jakob Disease in humans. About One in a million humans in the United States may contract the disease.

What would you do if you were stranded in a lifeboat at sea? Draw straws? It is called the "Lifeboat Strategy."

Cannibalism is not addressed by law in the United States. It is, however, covered by the criminal statutes regarding "corpse abuse." In Kentucky, corpse abuse is addressed in KRS 525.120, entitled "Abuse of Corpse," partially reprinted below:

> (1) A person is guilty of abuse of a corpse when, except as authorized by law, he intentionally treats a corpse in a way that would outrage ordinary family sensibilities ...
> (2) Abuse of a corpse is a Class A misdemeanor, unless the act attempted or committed involved sexual intercourse or deviate sexual intercourse with the corpse or the deliberate failure to prepare, bury, or cremate a corpse after the acceptance of remuneration into any contract negotiated, in which it is a Class D Felony.

For a discussion of Ohio's corpse abuse statute, see *Condon v. Wolfe*, 310 Fed. Appx. 807 (6th Cir. 2009). The famous 1884 English criminal case, *Regina v. Dudley and Stevens*, is perhaps the most instructive legal and cultural case published on the matter.

---

52 **Cassava (*Manihot esculenta*) is a South American carbohydrate-rich tuber.** Yet it is little known in the United States. We see it here mostly in the form of tapioca. Yet it is, after rice and corn, the third greatest source of carbohydrates in the tropical nations. More than 900 million people rely on it. And it is poisonous. Cyanide. *Time Magazine* labeled it as one of "the 10 most dangerous foods." Cassava must be significantly processed to remove the toxins. Then there is the delicacy of sliced fugu (blowfish). More cyanide. Yet 10,000 tons are eaten annually. Yikes!

---

53 **The great diversity of the natural potato stock can be seen at the Potato Park (Parque de la Papa) in Peru.** It is an indigenous Bio-cultural Heritage Area (IBCHA) where 600 varieties of potatoes grow. The Peruvian International Potato Center is even studying how to grow potatoes on Mars!

---

54 **A 46-mile long gash through the mountains, 6 miles wide at its narrowest,** where one could walk along a deep fault line past snow-capped mountains 13,000 feet above her.

---

55 **Thirty years ago, California produced 40% of the world's raisins.** Prices have dropped 25% from their 10-year average. As a result, only 165,000 acres are currently in raisins in California's Central Valley. In 2000, there were 280,000 acres in raisins. Many farmers are tearing out the grapevines and replacing them with almond trees at a cost of $7,000 per acre. Turkey and Iran are also significant producers of raisins, with Turkey expected to be the number one producer in the world.

Horne v. United States Department of Agriculture, No. 12-123, 569 U.S. _______ (2013), and No. 14-275, 576 U.S. _________ (2015) are cases dealing with the USDA's National Raisin Reserve. The Reserve is an outgrowth of Depression Era price collapse in the raisin industry. Congress authorized the USDA to obtain excess raisin production so as to buoy market prices. The plaintiff barred the government from gathering his raisin crop, thus engendering a legal review under the "Takings Clause" of the Fifth Amendment to the Constitution.

Writing for the majority, Chief Justice Roberts stated, in part:

> ...The Fifth Amendment applies to personal property as well as real property. The Governmenthas a categorical duty to pay just compensation when it takes your car, just as when it takes your home. ...
>
> This principle, dating back as far as Magna Carta, was codified in the Takings Clause in part because of property appropriations by both sides during the Revolutionary War. This Court has noted that an owner of personal property may expect that new regulation of the use of property could "render his property economically worthless. ... But there is still a "longstanding distinction" between regulations concerning the use of property and government acquisition

> of property. ... When it comes to physical appropriations, people do not expect their property, real or personal, to be actually occupied or taken away. ... The reserve requirement imposed by the Raisin Committee is a clear physical taking. Actual raisins are transferred from the growers to the Government. Title to the raisins passes to the Raison Committee. The Committee disposes of those raisins as it wishes, to promote the purposes of the raisin marketing order. The Government's formal demand that the Hornes turn over a percentage of their raisin crop without charge, for the Government's control and use, is of such a unique character that it is taking without regard to other factors that a court might ordinarily examine. (Horne at _________).

---

56 **Triangular trade can apply to a number of transoceanic shipping schemes.** One of the most infamous involved the "Middle Passage," carrying African slaves packed in the holds of sailing ships headed for the Caribbean Islands. It began in various forms in the 1500s, extending into the 1800s. Here is one way it worked in regard to sugarcane (*Saccharum sp.*). Sugarcane is native to Southeast Asia. It was imported into the Caribbean Islands by Christopher Columbus, where it flourished. Sugarcane cultivation is labor-intensive, however, and disease reduced the number of Native Americans available to work on the sugarcane plantations. Soon, European traders were sailing the first leg of the Triangular Trade, bringing manufactured goods to West Africa where they purchased slaves for the next leg to the Caribbean Islands, the Middle Passage.

In the Caribbean, African slaves grew and processed the sugarcane into refined sugar. Because of the nature of the trade winds, the next leg of the trade brought the products of sugarcane processing, sugar and molasses, to New England. There, molasses would be made into rum. These ships would then carry rum, sugar, tobacco, cotton, and finished goods back to Europe or directly to Africa.

Rum was made from molasses. Molasses was the concentrated dregs left after the sugar was removed from sugarcane syrup. It became a product to which value could be added through fermentation. A lower weight good with concentrated alcohol and enhanced profitability. This rum was prized in the Americas, Europe, and Africa as an essential trade ingredient. African slaves, Caribbean molasses, New England rum. A profitable, punishing example of the Triangular Trade.

The transatlantic slave trade, including that to Brazil, lasted approximately 366 years. The trade in North American slaves lasted approximately 250 years. There were as many as 12.5 million Africans enslaved and transported to the Americas. Twelve generations of families. In 1750, one male slave was valued at six ounces of gold; or two muskets; or forty pounds of gunpowder; or 8.5 gallons of Danish brandy; or twenty cowrie shells. In 1860, North American slaves were valued at approximately $3 billion in today's dollars. In 1850, one plantation owned 1,092 slaves, and produced four million pounds of rice each year. The rice was sent to the Caribbean to feed slaves working on the sugar plantations. One slave was expected to tend five acres and produce two thousand pounds of rice each year. (Civil Rights Museum Exhibits, Memphis, Tennessee, 2017).

In the 1760s, one slave was expected to produce 200 pounds of sugar, or 3.5 "hogs heads" per year on the Caribbean sugar plantations. Molasses was the residue of sugar processing. 110 gallons of molasses were converted in New England to 73 gallons of rum. Rum would then be transported to Africa where it would be used to purchase more slaves. In

1767, slaves could be bought for the following trade goods: one male for 130 gallons of rum; one female for 110 gallons of rum; one young female for 80 gallons of rum. (Civil Rights Museum Exhibits, Memphis, Tennessee, 2017).

---

57 **Corn, what can I say about it? Everything! That's the problem.** There is so much to say. Corn is everywhere. All throughout the U.S., all throughout the globe. People of the corn we truly are. And corn is color of the sun. Really, the sun. A child of that fierce fusion of hydrogen into helium. A sponge of solar chaos caught in a smooth yellow kernel of life. How wonderful. And it pops! But far too much of a journey to even begin here. You take the first steps of research and cultural exploration. I will catch up with you later.

---

58 **Agricultural data is telling about the human condition and its manifestations.** 75% of all agricultural lands support animal husbandry and 35% of our agricultural crops go to feed animals. The World Food Programme (United Nations) estimates that 795 million people globally are undernourished, primarily in Asia. In Sub Saharan, Africa, 25% of the population is underfed. This is so even though the global production of food increased approximately 30% from 1985 to 2005. Meat production is caloric intensive. In the United States, to grow one calorie of meat tissue, farmers utilize 29 calories of feed. And, generally, cattle require 20 times more water per calorie than grains for human consumption.

---

[59] **Bovines (family *Bovidae*) are a large group of ungulates (hooved mammals, formerly the order *Ungulata*).** Ungulates are now classified in two orders, *Artiodactyla* (even-toed, cloven hoof) and *Persissodactyla* (odd-toed). Artiodactyls include cattle, deer, pigs, camels, sheep, dolphins, whales and hippos. Persissodactyls include horses, tapirs and rhinos. The Bovidae family is comprised of Artiodactyls with 60 genera and 143 species, including cattle, sheep, bison, water buffalo, impalas, and goats. Deer are a member of another ruminant (see below) family of artiodactyls, *Cervidae*. Bovids have horns that are not annually shed, like deer. Bovid horns continue to grow year after year.

Another important characteristic of some bovids involves digestion. They are called ruminants. Ruminants "chew cud," hence their name, which is derived from the Latin word for chewing. They chew cud because they can. They have four stomachs and can regurgitate vegetation mass to re-chew and further breakdown the plant cellulose that bacteria is fermenting in their first two stomachs, the rumen and reticulum. The third stomach is the omasum, and the fourth, like a normal singular stomach, is the abomasum. This is "foregut" digestion. We don't have the stomachs or the bacteria to do it. It is a highly specialized system.

There are other animals that house bacteria and can eat and digest plant tissue. They are "hindgut" digesters where the bacteria necessary to ferment and breakdown cellulose lives in the ceacum, an enlarged area between the small and large intestines. These animals include elephants (order *Proboscidea*), and the bovid horses, and rhinos. Hindgut digestion is much less efficient than foregut digestion. But the ability to digest plant tissue increases the size of the candy store. And I won't even tell you how rabbits and termites go about it. You might lose your appetite!

The foregut, hindgut dichotomy has also influenced the evolution of plant-animal relationships and the dispersal of seeds. Some plants have evolved larger, bitter tasting seeds designed to deter crushing and pass through the digestive tract of hindgut fermenters. The seeds are then deposited in dung gardens throughout the land. The bitter toxins may offer pest protection but will also deter large animals from chewing the large seeds. Those seeds become passengers to faraway places on a lumbering bus. Not so for their ruminants cousins that are compelled to chew and rechew their cuds, crushing any large seeds. And the passages between stomachs are smaller in ruminants. Ruminants must also protect the alkaline environment in their foregut's digestion chambers. Too much fruit, sugar, or carbohydrates and the foregut acidifies. Not good for ruminants. And the bacteria will eat up all of the available sugar themselves!

By the way, microbes in those fore-stomachs produce a complete range of amino acids, as well as all the B-vitamins.

---

[60] **CAFOs are defined by the EPA as large, medium, and small.** 1,000 or more cattle are large CAFOs; 300 or fewer, small. For swine greater than 55 pounds in weight, 2,500 constitutes a large CAFO and less than 750, a small one. Less than 55 pounds? Well, it takes more than 10,000 swine to become a large CAFO in that case. Turkeys? 55,000 will get you in the large category, as will 125,000 non-laying chickens. See 40 CFR 122.23 for EPA's permit regulations regarding "Concentrated Animal Feeding Operations." 40 CFR 122.24 deals with "Concentrated Aquatic Animal Production Facilities." What constitutes an aquatic animal is described in Appendix C to Part 122.

Now for some data. This information gives perspective to the CAFO dialog.

In the United States in 1950, chicken consumption per person per year was 21 pounds. In 2000, it had risen to 77 pounds. In 1950, about 616,000 chickens were killed for meat. In 2000, it was approximately 8 billion. Yet egg consumption has declined per person; 380 eggs in 1950 and 247 per person in 2000.

In 1950, 65 pounds of pork was consumed per person. In 2007, it was 51 pounds. And the number of hog operations declined (and concentrated) from over three million to approximately 65,500 during that same time period.

Beef consumption annually per person increased from 44 pounds to 67 pounds from 1950 to 2000. (Imhoff, 2010).

According to the Humane Society, in the United States 1.8 billion animals were slaughtered in the United States in 1960. In 2014, they calculated that 9.1 billion animals were killed. (The Humane Society, "Farm Animal Statistics: Slaughter Totals;" www.humanesociety.org).

In 1958, the Humane Slaughter Act was passed. It has had three significant amendments. (7 U.S.C. § 1901 et seq.). The Humane Slaughter Act applies to cattle and pigs but not poultry. That exclusion was the subject of a lawsuit by the Humane Society and others against the Secretary of the United States Department of Agriculture. (*Levine v. Vilsack*, 587 F. 3d 986, 9th Cir. 2009). The plaintiffs sought to include poultry as "other livestock" under the Humane Slaughter Act. The 9th Circuit upheld the lower court's dismissal of the action, stating, "Because Levine's alleged injuries are not redressable by way of this lawsuit, there is a lack of standing to proceed with this action." (*Levine*, Supra, at P. 997). The "Findings and Declaration of Policy" in the Humane Slaughter Act are set out below:

> The Congress finds that the use of humane methods in the slaughter of livestock prevents needless suffering; results in safer and better working conditions for persons engaged in the slaughtering industry; brings about improvement of products and economies in slaughtering operations; and produces other benefits for producers, processors, and consumers which tend to expedite an orderly flow of livestock and livestock products in interstate and foreign commerce. It is therefore declared to be the policy of the United States that the slaughtering of livestock and the handling of livestock in connection with slaughter shall be carried out by humane methods. (7 U.S. Code § 1901)

Milk has a wonderful history, and its various products are mainstays of many diets. We have already discussed the fact that as adults, many humans lose the enzyme that was necessary to process lactose (milk sugar) in their mother's milk. That gene in many people of the world turns off the lactase enzyme production as the child matures. Only in dairy-dominated cultures has this gene been selected over time to continue the production of the enzyme in adulthood. Of course "soured" milk, where bacteria have turned milk sugar into lactic acid resulting in cheese and yogurts, is another subject.

Milk dairies have been big business in America. There may have been 650,000 American dairies in 1970, perhaps 65,000 (including concentrated operations) in 2000. Yet there has been an increase in milk cows to 9.2 million head producing 189,320 million pounds of milk in 2009. Milking operations have become larger with more milk produced per cow.

In 2015, total cheese production in the United States was 11.8 billion pounds, Wisconsin producing more than 26% of that total. Mozzarella, Cheddar, and American were the most produced, in that order. Butter production was 1.86 billion pounds. Yogurt, 4.7 billion pounds, along with 865 million gallons of regular ice cream.

---

[61] **Food figures prominently in the Bible.** For example, take Luke 9:13-17:

> 13. But He said unto them, Give ye them to eat. And they said, We have no more but five loaves and two fishes; except we should go and buy meat for all this people.
>
> 14. For they were about five thousand men. And He said to His disciples, Make them sit down by fifties in a company.
>
> 15. And they did so, and made them all sit down.
>
> 16. Then He took the five loaves and the two fishes, a n d looking up to heaven, He blessed them, and brake, and gave to the disciples to set before the multitude.
>
> 17. And they did eat, and were all filled; and there was taken up of fragments that remained to them twelve baskets.

See also Matthew 14:21; Mark 6:31-44; and John 6:5-15 (King James Version).

A second miraculous feeding of the masses is reported in Matthew 15:32-38 (King James Version). See also Mark 8: 1-9 (King James Version).

Water turning into wine is described in John 2:1-11 (King James Version). The catching of fish in Luke 5:4-10 and again in John 21:1-11 (King James Version).

There are also many descriptions of food described in the Quran. For example, see honey (Quran, 16:69), fish (Quran, 18:61-63), the olive (Quran, 24:35), pork (Quran, 2:173), the date (Quran, 19:23-26), milk (Quran, 16:66), and the fig

(Quran, 95:1). And the Old Testament contains many additional dietary instructions.

There is a significant debate in the United States concerning the meaning of the First Amendment of the United States Constitution. It is often referred to as the "Establishment Clause," and states: "Congress shall make no law respecting an establishment of religion, or prohibiting the free exercise thereof; ..."

There are many Supreme Court decisions involving the Establishment Clause of the constitution and the separation of church and state. *Town of Greece v. Galloway*, (2014) 134 S. Ct. 1811, 188 L. Ed 2d 835, is instructive of the issues and dynamics before the court. The case upheld prayer in school and Justice Alito, concurring with the majority opinion of the court, stated in part as follows:

> This brings me to my final point. I am troubled by the message that some readers may take from the principal dissent's rhetoric and its highly imaginative hypotheticals. For example, the principal dissent conjures up the image of a litigant awaiting trial who is asked by the presiding judge to rise for a Christian prayer, of an official at a polling place who conveys the expectation that citizens wishing to vote make the sign of the cross before casting their ballots, and of an immigrant seeking naturalization who is asked to bow her head and recite a Christian prayer. Although I do not suggest that the implication is intentional, I am concerned that at least some readers will take these hypotheticals as a warning that this is where today's decision leads to a country in which religious minorities are denied the equal benefits of citizenship.
>
> Nothing could be further from the truth. All that the Court does today is to allow a town to follow a practice that we

> have previously held is permissible for Congress and state legislatures.
>
> In seeming to suggest otherwise, the principal dissent goes far astray. (572 U.S. __________, (2014); Slip Opinion, pg. 13).

In dissent, Justice Kagan wrote, in part, as follows:

> For me, that remarkable guarantee means at least this much: When the citizens of this country approach their government, they do so only as Americans, not as members of one faith or another. And that means that even in a partly legislative body, they should not confront government-sponsored worship that divides them along religious lines. I believe, for all the reasons I have given, that the Town of Greece betrayed that promise. I therefore respectfully dissent from the Court's decision. (572. U.S. _____, (2014); Slip Opinion, pg. 25).

---

[62] **See Finkelstein and Silberman, 2001, pg. 119.**

A restriction on swine is found in both the Old Testament and the Quran.

> And the swine, though he divide the hoof, and becloven-footed, yet he cheweth not the cud; he is unclean to you. (Leviticus 11:7, King James Version).

> He has only forbidden to you dead animals, blood, the flesh of swine, and that which has been dedicated to other than Allah. (Quran, Surah 2:173).

I would be negligent in my reporting duties if I did not point out a section from the Old Testament Book of Genesis that is often referred to as the "Dominion Clause." It is often interpreted as giving humans sway over all plant and animal life with the power to use them, as well as the Earth itself, to their needs and interests. Here is the King James Version:

> Then God said, let us make man in our image, after our likeness, and let them have dominion over the fish in the sea, and the fowl of the air, and over the cattle, and over all the Earth, and over every creeping thing that creepeth upon the Earth. (Genesis 1:26, King James Version).

> The Quran has similar language: Do ye not see that Allah has subjected to your (use) all things in the heavens and on Earth, and has made his bounties flow to you in exceeding measure, both seen and unseen? (Quran, Surah 31:20).

Thus much has been done to this planet and its life forms using religion as support for the authority of human actions.

---

[63] **In 2005, India's Supreme Court overruled the Gujarat High Court's** determination that such a ban was unconstitutional. The law in question was the Bombay Animal Preservation Act of 1994. *State of Gujarat v. Mirzapur Moti Kureshi Kassab Jamat Ops.*, 2005, Supreme Court of India, Appeal (civil) 4937-4940 of 1998.

The Court's decision was based on Article 48 in the Constitution of India 1949 which states:

> Organisation of agriculture and animal husbandry: The State shall endeavour to organise agriculture and animal husbandry on modern and scientific lines and shall, in particular, take steps for preserving and improving the breeds, and prohibiting the slaughter, of cows and calves and other milch and draught cattle [sic].

In 2016, an owner of a roadside food stall was arrested for selling meat in a rice and vegetable dish in the Indian state of

Haryana. He faces a maximum penalty of ten years for the slaughter and five years for selling the meat. (Sorge, 2016).

---

[64] **Native Americans called the tree, Pauame, the French, Sassafras.** Nicolas Monardes of Seville in 1574 reported that it was a tree "... of greate vertues, and great excellencies, that thei heale there with greevous and variable diseases [sic] "(Peattie, 1948). Early exploration of America included the search and export of sassafras.

It is the root bark that contains aromatic essential oils. Such aromatic trees were uncommon in Europe. Sassafras was native to North America and Asia. Its use as a "cure-all" has faded, but it is still popular in foods, soaps, and perfumes. It is also used as a demulcent (soothing application for irritation) and an emollient (skin moisturizing).

---

[65] **Freshwater mussels also have a place in the industrial history of our culture.** One hundred years ago, there were sixty factories along the Mississippi River with machines that punched out plug-sized sections from thick walled freshwater mussels such as the pistol grip (*Tritogonia verrucosa*). They were then milled for use as buttons. By 1930, over-harvesting caused the industry to decline and substitutes to be found for buttons.

By 1950, the industry was back, but not for buttons. Kokichi Mikimoto of Japan discovered that bits of freshwater mussels used as starter material caused cultured marine oysters to produce beautiful pearls. Chances are that the pearl on your dresser is winking at you. It may have been your neighbor and started out as a freshwater mussel right here in Kentucky.

"Brailling" is fully regulated in Kentucky. That is how the mussels are harvested. Blunt hooks are attached to chains and allowed to drag across mussel beds. The shells close on the hooks and are pulled up with the chains. You can see this operation in the 1955 movie, "The Kentuckian." Burt Lancaster is in search of a freshwater pearl to make his fortune.

In Kentucky, a license is needed to take mussels. KRS 150-520 and KRS 150-525. See also 301 KAR 1:085.

---

[66] **Kentucky is a state of small farmers.** About 76,500 of them, averaging 170 acres per farm. 51% of Kentucky land is farmed. 57% of these farms averaged less than $10,000 in annual sales. Total receipts from Kentucky farms in 2013 were approximately $5.7 billion. In 2013, Kentucky ranked 16th in farm income, but second in tobacco production. The top four agricultural exports in Kentucky for 2013 were: 1) soybean; 2) other livestock products; 3) wheat; and 4) poultry products. Shelby County is the 4th in number of farms, and Barren County is first.

Poultry production is the top Kentucky agricultural product. 1,274 Kentucky farms raised broilers and meat chickens in 2012. The top five counties are in Western Kentucky. There were more than one million head of cattle in Kentucky in 2015, more than any state east of the Mississippi River. Kentucky has about 714 dairy farms, averaging 88 milk cows per dairy. That's 115 million gallons of milk in 2014. In swine, Kentucky ranks 21st, with 325,000 head in 2014.

Kentucky is first among states in horse sales and has 35,000 horse-based operations. There are approximately 242,000 horses in Kentucky. In 2015, there were 65,700 goats and 48,000 sheep in Kentucky. Shelby County is fourth in the state for the number of sheep.

Kentucky is 14th in state rankings for corn. Corn is second only to soybeans in production in the state. In 2014, Kentuckians harvested 1,300,000 acres, producing almost 225 million bushels of corn (about 158 bushels per acre). The counties with the biggest production are in Western Kentucky.

Soybeans are the number one agricultural product in the state. 1,750,000 acres harvested in 2014, averaging 48 bushels per acre. Again, Western Kentucky led in production. (Kentucky Farm Bureau, 2015)

---

[67] **The Federal Environmental Protection Agency has recently been the focus of political criticism.** Some of that criticism is based on constitutional arguments, including the Tenth Amendment and the "Interstate Commerce Clause." Those constitutional provisions state as follows:

> The powers not delegated to the United States by the Constitution, nor prohibited by it to the states, are reserved to the states respectively, or to the people. (The United States Constitution, Amendment X).

> The Congress shall have Power ... to regulate Commerce with foreign Nations, and among the several states, and with the Indian tribes, (The United States Constitution, Article 1, Section 8, Clause 3).

But see also the "Necessary and Proper Clause" of the United States Constitution:

> To make all laws which shall be necessary and proper for carrying into Execution the foregoing Powers, and all other Powers vested by this Constitution in the Government of the United States, or in any Department or Office thereof.

(The United States Constitution, Article 1, Section 8, Clause 18).

Chief Justice Marshall of the United States Supreme Court wrote the seminal case on the "Necessary and Proper Clause" in 1819. It is entitled *McCulloch v. Maryland*, 17 U.S. 316 (1819). Go ahead and Google the citation. The Supreme Court posts its Decisions to the web. Fun reading, especially in times of change!

---

[68] **33 U.S.C. § 407. Also known as "The Refuse Act," this provision is the oldest statute dealing with environmental law in the United States.** It states, in part:

> It shall not be lawful to throw, discharge, or deposit, or cause, suffer, or procure to be thrown, discharged, or deposited either from or out of any ship, barge, or other floating craft of any kind, or from the shore, wharf, manufacturing establishment, or mill of any kind, any refuse matter of any kind or description whatever other than that flowing from streets and sewers and passing therefrom in a liquid state, into any navigable water of the United States, or into any tributary of any navigable water from which the same shall float or be washed into such navigable water; and it shall not be lawful to deposit, or cause, suffer, or procure to be deposited material of any kind in any place on the bank of any navigable water, or on the bank of any tributary of any navigable water, where the same shall be liable to be washed into such navigable water, either by ordinary or high tides, or by storms or floods, or otherwise, whereby navigation shall or may be impeded or obstructed. ...
>
> (33 U.S.C. § 407).

Notice the use of the term, "navigable water." Much litigation has sought to clarify the meaning of that term! (See *Rapanos v. United States*, 547 U.S. 715 (2006)).

A sharp focus of President Donald Trump's Administration is the repeal or modification of the definition of "Waters of the United States" under the Clean Water Act. The Final Rule at issue was published in the *Federal Register* on June 29, 2015. (80 F.R. 37054):

> Congress enacted the CWA 'to restore and maintain the chemical, physical, and biological integrity of the Nation's waters,' section 101(a), and to complement statutes that protect the navigability of waters, suchas the Rivers and Harbors Act. 33 U.S.C. 401, 403, 404, 407. The CWA is the nation's single most important statute for protecting America's clean water against pollution, degradation, and destruction. To provide that protection, the Supreme Court has consistently agreed that the geographic scope of the CWA reaches beyond waters that are navigable in fact. Peer-reviewed science and practical experience demonstrate that upstream waters, including headwaters and wetlands, significantly affect the chemical, physical, and biological integrity of downstream waters byplaying a crucial role in controlling sediment,filtering pollutants, reducing flooding, providing habitat for fish and other aquatic wildlife, and many other vital chemical, physical, and biological processes. ... (80 F.R. 37055)

The United States Congress has also utilized the Congressional Review Act of 1996 (5 U.S.C. § 8) to overrule the Office of Surface Mining Reclamation and Enforcement's "Stream Protection Rule." That action is subject to President Trump's signature.

---

[69] **FWPCA is now codified as part of the Clean Water Act and is codified at 33 U.S.C. §§ 1251 et seq.** The FWPCA was based on a "command and control" template for protecting human health and the environment, first promulgated under the Clean Air Act of 1970. (42 U.S.C. §§ 7401-7671Q). It would become the model for ten more years of federal environmental legislation.

---

[70] **The largest underground aquifer in the United States is in the Ogallala Aquifer, stretching from Nebraska to Texas.** Yet many of the world's aquifers are subsiding due to groundwater withdrawal. (Peabody, 2016).

---

[71] **"Food stamps and nutrition, $756 billion;** crop insurance $89.8 billion; conservation $56 billion; commodity programs $44.4 billion; everything else $8.2 billion." (Plumer, 2014).

---

[72] **"The human population of the Earth at the beginning of the Christian Era (CE) was approximately 300 million.** By 1500 CE, it was about 500 million people. Around 1800, it was approximately 1.0 billion. In 1950, it had climbed to more than 2.5 billion. In 2014, the world's population was greater than 7.1 billion people. Reasonable projections of the Earth's human population in 2050 are at 9.6 billion people.

The number of Native Americans inhabiting Kentucky prior to 1500 is unknown. Archaeological evidence, however, indicates that the Commonwealth was thoroughly settled at different times prior to European contact.

Regardless, after obtaining firearms from the Dutch during the "Beaver Wars" of the 1600s, the Iroquois drove out the Shawnee from Ohio and Kentucky. As a result, land explorer Christopher Gist, in 1751, found only one Native American settlement in Kentucky, that of forty dwellings in lower Shawneetown in South Shore, Kentucky opposite modern Portsmouth, Ohio and the mouth of the Scioto River. That and Eskippakithiki (Indian old fields) near Winchester, Kentucky, may have been the only permanent Native American village at the time of European colonization of the Commonwealth." (Van Stockum, Jr., Fall/Winter 2015).

This was not to be typical of population growth in the world as a whole. It did not take long to understand the potential of human reproduction.

> I think I may fairly make two postulata. First, that food is necessary to the existence of man. Secondly, that passion between the sexes is necessary and will remain nearly in its present state ... but toward the extinction of the passion between the sexes, no progress whatever has hitherto been made ...
>
> Population, when unchecked, increases in a geometric ratio. Sustenance increases only in an arithmetical ratio... I see no way by which man can escape from the weight of this law which pervades all of animated nature. (Malthus, 1798). (Van Stockum, Jr., footnote 10, Fall/Winter 2015).

> Americans are beginning to realize that the underdeveloped countries of the world face an inevitable population-food crisis. Each year food production falls a bit further behind burgeoning population growth, and people go to bed a little bit hungrier. While there are temporary or local reversals of this trend, it now seems inevitable that it will continue to

> its logical conclusion: mass starvation. The rich may continue to get richer, but the more numerous poor are going to get poorer. Of these poor, a minimum of ten million people, most of them children, will starve to death during each year of the 1970s. But this is a mere handful compared to the numbers that will be starving before the end of the century. And it is now too late to take action to save many of those people. (Ehrlich, 1968).
>
> (Van Stockum, Jr., Fall/Winter 2015).

That did not happen, of course, but there are a mighty lot of people on this planet, and some countries have taken action.

China instituted a "one child per family" policy in 1979. Using a combination of benefits and penalties, the country has tried to control population growth. This policy has been difficult to administer and has had many exceptions that applied differently in urban versus rural areas. The policy has also possibly exacerbated a cultural preference for male offspring. There is now an excess of 33 million more men than women in China. In 2013, these policies were relaxed allowing, along with other changes, two children when one parent is an only child.

Japan, on the other hand, is actively seeking to increase its birthrate. As the Japanese population has aged, the country has been unable to sustain its numbers. The government seeks, through economic incentives, to maintain its population at 100 million people. Its current population is approximately 127 million and is expected to fall as low as 87 million by 2060 if there is no intervention. Japan is a homogeneous culture with immigration generally disfavored as a method of increasing population.

India is expected to be the world's most populous nation in 2030, with 1.5 billion people. The current population is approximately 1.25 billion. Female sterilization has been a voluntary method of birth control in India (and China also). Sterilization is often accompanied with payment by the state. India also has a surplus of male children due to selective abortion or the practice of infanticide. In India, cultural bias has resulted in few male vasectomies.

---

73 **According to the results of the *2015 Revision*,** the world population reached 7.3 billion as of mid-2015 (table 1), implying that the world has added approximately one billion people in the span of the last twelve years. Sixty per cent of the global population lives in Asia (4.4 billion), 16 per cent in Africa (1.2 billion), 10 per cent in Europe (738 million), 9 per cent in Latin America and the Caribbean (634 million), and the remaining 5 per cent in Northern America (358 million) and Oceania (39 million). China (1.4 billion) and India (1.3 billion) remain the two largest countries of of the world both with more than 1 billion people, representing 19 and 18 per cent of the world's population, respectively. ... Currently, the world population continues to grow though more slowly than in the recent past. Ten years ago, world population was growing by 1.24 per cent per year.

Today, it is growing by 1.18 per cent per year, or approximately an additional 83 million people annually. The world population is projected to increase by more than one billion people within the next 15 years, reaching 8.5 billion in 2030, and to increase further to 9.7 billion in 2050 and 11.2 billion by 2100 ... (United Nations, 2015).

---

74 **In January 2017, the National Oceanic and Atmospheric Administration of the United States**

Department of Commerce issued a report entitled "Global and Regional Sea Level Rise Scenarios for the United States."

> We recommend a revised extreme upper-bound scenario for GMSL [Global Mean Sea Level] rise of 2.5 m [8.2 feet] by the year 2100 ... These scenario-based RSL values fill a major gap in climate information needed to support a wide range of assessment, planning, and and decision-making processes. GMSL was adjusted to account for key factors important at regional scales, including: 1) shifts in oceanographic factors such as circulation patterns; 2) changes in the Earth's gravitational field and rotation, and the flexure of the crust and upper mantle, due to melting of land-based ice; and 3) vertical land movement (VLM; subsidence or uplift) due to glacial isostatic adjustment (GIA, which also changes Earth's gravitational field and rotation, as well as the overall shape of the ocean basin), sediment compaction, groundwater and fossil fuel withdrawals, and other nonclimatic factors....(NOAA, 2017).

The International Panel on Climate Change had this to report:

> Sea level rise projections over the next century vary considerably, with the high-end scenarios yielding a rise of up to 6 or 7 feet by 2100. About 7 to 8 million people in the U.S. live within 6 feet of the local high tide line, and storm surge can extend flooding far beyond the high tide line, as witnessed in Superstorm Sandy. (IPCC, 2013).

---

[75] **Obliquity refers to the planet's tilt.** It is currently at an angle of 23.5 degrees but changes over a 44,000-year cycle. Eccentricity involves the circuit of the Earth's orbit around the sun. That circuit bulges slightly on a 96,000-year cycle.

Precession relates to the spinning of Earth. Like the top that is spun with a wound string, Earth wobbles in its spin. Its axis migrates over a 23,000-year cycle, describing a circle in the north sky. That's why 2,000 years ago sailors were not navigating by the North Star. The celestial North Pole then lay in darkness.

---

76 **The amount of energy released can equal all of electricity that would be used on Earth for a million years.** Giant prominences flare out from the sun at the distances equal to that from Earth to the moon. And the sun's temperature is astoundingly hot. If a pin was heated on Earth to the temperature in the center of the sun, it would instantly incinerate anything within 60 miles of it. That's 154 million degrees kelvin.

---

77 **"Warming of the climate system is unequivocal,** and since the 1950s, many of the observed changes are unprecedented over decades to millennia. The atmosphere and ocean have warmed, the amounts of snow and ice have diminished, sea level has risen, and the concentrations of greenhouse gases have increased. ... Each of the last three decades has been successively warmer at the Earth's surface than any preceding decade since 1850. In the Northern Hemisphere, 1983-2012 was *likely* the warmest 30-year period of the last 1400 years ...

The atmospheric concentrations of carbon dioxide, methane, and nitrous oxide have increased to levels unprecedented in at least the last 800,000 years. Carbon dioxide concentrations have increased by 40% since pre-industrial times, primarily from fossil fuel emissions and secondarily from net land use change emissions. The ocean has absorbed about 30% of the emitted anthropogenic carbon dioxide, causing ocean acidification. ... It is extremely likely that human influence has been the dominant cause of the observed warming since the mid-20th century." (IPCC, 2013)

---

78 **"Climate scientists agree: climate change is happening here and now.** Based on well-established evidence, about 97% of climate scientists have concluded that human-caused climate change is happening ... The evidence is overwhelming: levels of greenhouse gases in the atmosphere are rising. Temperatures are going up. Springs are arriving earlier. Ice sheets are melting.

Sea level is rising. The patterns of rainfall and drought are changing. Heat waves are getting worse as is extreme precipitation. The oceans are acidifying. ... The oceans are absorbing much of the CO_2 that smokestacks and tailpipes emit into the atmosphere.

As a result, the oceans are rapidly acidifying, with early impacts on shelled organisms such as oysters already documented. The current acidification rate is likely the fastest in 300 million years." (AAAS, 2014)

---

79 **On February 24, 2017, the National Weather Service at the Louisville International Airport recorded temperatures of 79° F.** That is the highest temperature recorded there during the month of February over the 134 years data has been collected. In 1883, 1889, and 1932, temperatures reached 78° F. The months of January and early February 2017 have been 6.7° F and 5.9° F warmer than the average temperature for these time periods. The warmest winter in

Kentucky was in 1931-1932 when it averaged 44.2° F across the state.

One report suggests that the reduction of air pollution, especially aerosols, has allowed for the penetration of more sunlight over the industrialized nations. The authors calculate that this "Solar Brightening" is a significant factor in the increase in American corn yields of 27% over the past thirty years. It is probable that the increase in atmospheric carbon dioxide accounts for as much as 6% of that total (Tollenaar, 2017).

Jim Ellis is an engineer and a long-time farmer in Shelby County, Kentucky. He now farms with his son, Seth, the third generation of his family farming legacy. Here is what Jim says:

> We have experienced significant weather events in our corn and soybean crops. The summer of 2012 was so hot that the corn pollen died and corn yields were greatly reduced. In the following crop years the yields have been record highs. Never before has the climate been so warm and wet as it was in these recent growing seasons.
> (Ellis, personal communication, 2017).

---

[80] **Because weather station locations and measurement practices change over time,** there are uncertainties in the interpretation of specific year-to-year global mean temperature differences. However, even taking this into account, NASA estimates 2016 was the warmest year with greater than 95 percent certainty. '2016 is remarkably the third record year in a row in this series,' said GISS Director

Gavin Schmidt. 'We don't expect record years every year, but the ongoing long-term warming trend is clear.' The planet's average surface temperature has risen about 2.0 degrees Fahrenheit (1.1 degrees Celsius) since the late 19th century, a change driven largely by increased carbon dioxide and other human-made emmissions into the atmosphere. Most of the warming occurred in the past 35 years, with 16 of the 17 warmest years on record occurring since 2001.

Not only was 2016 the warmest year on record, but eight of the 12 months that make up the year, from January through September, with the exception of June, were the warmest on record. ...

("NASA, NOAA Data Show 2016 Warmest
Year On Record Globally," Release 17-006,
January 18, 2017).

---

81 **A search of definitions within the United States Environmental Protection Agency (EPA) retrieves the following:**

[1]Genetic Engineering

Definition: A process of inserting new genetic information into existing cells in order to modify a specific organism for the purpose of changing one of its characteristics.

[2]Genetic Engineering

Definition 1: Directed modification of the gene complement of a living organism by such techniques as altering the DNA, substituting genetic material by means of a virus, transplanting whole nuclei, transplanting cell hybrids, etc. [NIH Genetics Home Reference Glossary]

Definition 2: Procedure used to join together DNA segments in a cell-free system (an environment outside a cell or organism). Under appropriate conditions, a recombinant DNA molecule can enter a cell and replicate there, either autonomously or after it has become integrated into a cellular chromosome. [NIH Genetics Home Reference Glossary]

[3]*Definition 3:* Genetic engineering is the process of using recombinant DNA (rDNA) technology to alter the genetic makeup of an organism. Traditionally, humans have manipulated genomes indirectly by controlling breeding and selecting offspring with desired traits. Genetic engineering involves the direct manipulation of one or more genes. Most often, a gene from another species is added to an organism's genome to give it a desired phenotype. [NIH Genetics Home Reference Glossary].

(aspub.epa.gov, "EPA terms and Acronyms")

---

[82] **Research by Paul Berg, Herbert Boyer, Stanley Cohen, and others.**

---

[83] **In 1981, the Nobel Prize in physiology or medicine went to Werner Arber, Hamilton Smith, and Daniel Nathans for the discovery of restriction enzymes.** The Nobel Committee stated in its press release as follows (in part):

> Restriction enzymes provide the 'chemical knives' to cut genes (=DNA) into defined fragments. These may then be used (1) to determine the order of genes on chromosomes,
>
> (2) to analyze the chemical structure of genes and of regions of DNA which regulate the function of genes, and (3) to create new combinations of genes. These techniques

open up new avenues to study the organization and expression of genes of higher animals and to solve basic problems in developmental biology. In medicine, increased knowledge in this area should help in the prevention and treatment of malformations hereditary diseases and cancer...

... Restriction enzymes are used as tools to dissect DNA into smaller defined fragments. These can be used to determine the order of genes on chromosomes, to analyze the chemical structure of genes and to recombine genes by chemical means. Most important restriction enzymes are used to analyze the function of regions of DNA which regulate gene expression...

---

[84] **In 1993, Kary Mullis gained the 1993 Nobel Prize in chemistry for the development of PCR.** The Nobel Committee stated in its press release as follows (in part):

The applications of Mullis' PCR method are already many. It is for example possible using simple equipment to multiply a given DNA segment from a complicated genetic material millions of times in a few hours, which is of very great significance for biochemical and genetic research. The method offers new possibilities particularly in medical diagnostics, and is used, for example, for discovering HIV virus or faulty genes in hereditary diseases.

Researchers can also produce DNA from animals that became extinct millions of years ago by using the PCR method on fossil material.

The genetic code programmed into the DNA Molecule determines the number and sequence of amino acids in a protein, and thus also the functional properties of the protein. ...Chemically, the genetic material of living organisms consists of DNA (deoxyribonucleic acid). DNA molecules consist of two very longstrands twisted around each other to form adouble helix. Each strand is formed of smaller molecules, nucleotides, that represent the letters of the genetic material. There are only four different letters, designated A, T, C and G. The two DNA strands are complementary, being held together by A - T and G - C bonds. It is only when the genetic code is to be read off e.g. for protein building in the cell that the two strands are separated. The genetic information in DNA exists as a long sentence of code words, each of which consist of 3 letters which can be combined in many different ways (e.g. CAG, ACT, GCC). Each three-letter code word can be translated by special components withinthe cell into one of the twenty amino acids that build up proteins. It is the proteins that are responsible for the functions of living cells, including their ability to function, among other things, as enzymes maintaining all the chemical reactions required for supporting life. The proteins' three-dimensional structure and hence their function is determined by the orderin which the various amino acids are linked together during protein synthesis. ...

---

[85] **It reproduces through an asexual budding process and is approved for use by the United States Food and Drug Administration (FDA) as a "Generally Recognized as Safe" (GRAS) organism.** (21. U.S.C. § 321(s); 21 CFR Part 182).

---

[86] **Glyphosate is the active ingredient in the Monsanto Company's registered herbicide "Roundup."** Many of the

GMOs listed for this footnote are considered "Roundup Ready." Glyphosate is a type of organic phosphorus compound. The chemical formula is $C_3H_8NO_5P$. It is effective by disrupting amino acid synthesis (tyrosine, tryptophan, and phenylalanine). It works on many grasses, woody plants, and other herbaceous growth that competes with agricultural production. It has enabled the use of "no till" farming, thereby reducing erosion. Although regulatory approval has been widespread, concerns exist as to the chemical's environmental fate, toxicity, and the evolution of glyphosate resistant weeds and pests.

---

[87] **CRISPR is the most recent "darling" of the genetic engineering world.** This powerful but yet relatively simple system is based on primitive microbial cell systems for identifying and disabling viruses. It was examined in a microbe, *Haloferax mediterrane*, living in extremely salty marshes in Spain. The system was investigated by Francisco Mojica at the University of Alicante, who published data in 1993 (Mojica et al. 1993). The molecular system has now been found in numerous microbes, including bacteria such as *Mycobacterium tuberculosis*, *Clostridium difficile*, and *Yersinia pestis*.

Three types of CRISPR systems have been described, with perhaps the Type II system in *Streptococcus thermophilus* being the simplest. Generally put, the CRISPR Associated Proteins (such as the RNA-guided nuclease CAS 9) seeks out and cleaves any DNA that matches the sequence of an invading virus. These sequences, over time, have been catalogued in the bacterium's genome as an immunity guide to destroy viral attackers with similar sequences. CAS 9, in association with CR-RNA (CRISPR RNA) and TRACR-RNA (trans-activating

CRISPR RNA) and G-RNA (Guide RNA) provide the acquired immunity now found in many bacteria and archean microbes.

Researchers modified the CRISPR system so as to provide guide RNA alignment with specific DNA sequences. Many have contributed to research in this area and significant litigation concerning patent rights has developed. Current litigation centers between the University of Berkeley and the Broad Institute of Harvard and MIT.

---

[88] **In addition to the BT toxins (Cry proteins),** agricultural crops have also been engineered to resist pests by proteins that inhibit insect metabolism involving protease, 2-amylase, lectins, cholesterol and tryptophan. For example, alfalfa, tobacco, and apple have received protease inhibitors.

---

[89] **The Monsanto Company, in association with the DOW Chemical Company, has produced genetically modified corn seed called "Smart Stax."** The corn is resistant to four herbicides (two from DOW - Hercules Xtra and Liberty Link; and two from Monsanto - Yieldgard VT and Roundup Ready 2). DOW lists eight insects that Smart Stax controls, including: 1) Western corn rootworm larvae; 2) Fall armyworm; 3) Corn ear worm; 4) Southwestern corn borer; 5) Black cutworm; 6) European corn borer; 7) Western bean cutworm; and 8) Northern corn rootworm larvae. These are caterpillars and larvae of moth and butterfly in the family *Lepipoptera*, class *Insecta*.

---

[90] **New York Times, April 13, 2016,** Andrew Pollack, "Acreage for Genetically Modified Crops Declined in 2015."

---

[91] **International Service for the Acquisition of Agri-Biotech Applications** (ISAAA, 2015).

---

[92] **India was an early entrant into GMO regulation** (see the "Rules of 1989, December 5, 1989, The Environmental Protection Act of 1986). The rules of 1989 set-up committees on recombinant DNA, institutional biosafety, and genetic engineering approval. Guidelines were published in 1990, 1994, and 1998. "The Seed Policy," 2002, was established dealing with trans genetic plants. In 2003, plant quarantine procedures were established and field trials regulated in 2008. (Rastogi, 2009).

---

[93] **35 U.S.C. § 101 states (in part):**

"Whoever invents or discovers any new and useful process, machine, manufacture, or composition of matter, or any new and useful improvement thereof, may obtain a patent therefor, subject to the conditions of this title. (35 U.S.C. § 101).

Chief Justice Warren Burger wrote the majority opinion for the court. In it, he made the following illustrative statements:

The grant or denial of patents on micro-organisms is not likely to put an end to genetic research or to its attendant risks. The large amount of research that has already occurred when no researcher had sure knowledge that patent protection would be available suggests that legislative or judicial fiat as to patentability will not deter the scientific mind from probing into the unknown any more than Canute could command the tides. Whether respondent's claims are patentable may determine whether research efforts are accelerated by the hope of reward or slowed by want of incentives, but that is all." (*Diamond*, 447 U.S. 318).

---

[94] **"Today's release of the 2017 Update to the Coordinated Framework for the Regulation of Biotechnology** represents the first time in 30 years that the Federal government has produced a comprehensive summary of the roles and responsibilities of the three principal regulatory agencies with respect to regulating biotechnology products. This update and the accompanying *National Strategy for Modernizing the Regulatory System for Biotechnology Products (Strategy)* offer the public a complete picture of robust and flexible regulatory structure that provides appropriate oversight for all products of modern biotechnology."

("Increasing the Transparency, Coordination, and Predictability of the Biotechnological Regulatory System," White House Blog, January 4, 2017).

---

[95] **"The policy of the United States Government** is to seek regulatory approaches that protect health and the environment while reducing regulatory burdens and avoiding unjustifiably inhibiting innovation, stigmatizing new technologies, or creating trade barriers. ... In some cases, uncertainty about agency jurisdiction, lack of predictability of timeframes for review, and other processes have imposed unnecessary costs and burdens on small and mid-sized companies and academics."

(Executive Summary," *National Strategy for Modernizing the Regulatory System for Biotechnology Products*, September 2016).

---

[96] **"While the current Federal regulatory system for biotechnology** products effectively protects health and the environment, advances in science and technology have altered the product landscape rapidly. ... directing the primary agencies that regulate the products of biotechnology, the U.S. Environmental Protection Agency (EPA), the U.S. Food and Drug Administration (FDA), and the U.S. Department of Agriculture (USDA), to update the Coordinated Framework for the Regulation of Biotechnology (Coordinated Framework) ..."

("Executive Summary," *Modernizing the Regulatory System for Biotechnology Products: Final Version of the 2017 Update to the Coordinated Framework for the Regulation of Biotechnology*).

---

[97] **"FDA is issuing this draft revised Guidance for Industry to clarify its approach to the regulation of intentionally altered genomic DNA in animals.** This guidance addresses animals whose genomes have been intentionally altered using modern molecular technologies, which may include random or targeted DNA sequence changes including nucleotide insertions, substitutions, or deletions, or other technologies that introduce specific changes to the genome of the animal.

This guidance applies to the intentionally altered genomic DNA in both the founder animal in which the initial alteration event occurred and the entire subsequent lineage of animals that contain the genomic alteration. ...

... More recently, new technologies have emerged that are intended to alter the genomes of various organisms, including animals. Some of these include the use of 'nucleases' or 'genome editing technologies' including engineered nuclease/nucleotide complexes such as zinc finger nucleases (ZFN), transcription activator-like effector nucleases (TALENs), and the clustered regulatory interspersed short palindromic repeats CRISPR) associated systems. These nucleases are intended to introduce alterations at specific sites in the genome, rather than the more random changes associated with rDNA technology. ..."

("Introduction and Background," *FDA Draft Guidance for Industry #187*)

---

[98] **Rats, bats, and saber-tooth cats**. There are 33 families, 481 genera, 2,277 species of rodents. The most numerous mammal group on Earth, 40% of the total. And that does not even include extinct species of which there may be five times as many. Rodents include rats, mice, squirrels, capybaras, porcupines, muskrats, beavers, chipmunks, hamsters, guinea pigs, chinchillas, gophers, and voles.

Bats are the only mammals that fly, excluding those few that "sail" on fleshy skin extensions. 20% of mammal species are bats. 18 families, 186 genera, and 1,240 species are currently extant. Only rodents have more species.

The protagonist at the beginning of this work is fortunate in his timing. Perhaps as recently as 10,000 years ago, the large saber-toothed cat, *Smilodon fatalis*, roamed this land. *Smilodon* is a true cat, but saber-tooth predators convergently evolved in other types of mammals.

Have you ever watched a cat chew through a tough piece of meat or crush a hard piece of meal on the side of its jaw?

The carnivorans and creodonts have large, specialized cutter teeth. They are called carnassial teeth. So the *Smilodon* had stabbing canine fangs, scissor-like cheek teeth, and crushing carnassials. Be glad that they are not still around!

XX.
Index

XXI.

RONALD R. VAN STOCKUM, JR.

Ronald R. Van Stockum, Jr. is a lawyer, teacher, biologist, writer, guitarist, and recently an actor living on his family's old farm in Shelbyville, Kentucky.

He has a B.S. in Biology from Santa Clara University, and an M.S. and PhD. in Biology from the University of Louisville. He also has his Juris Doctorate in Law from the University of Louisville and practices law from offices in Shelbyville, Kentucky. He concentrates his legal practice in environmental law, and his biologic research in historical phytogeography.

Dr. Van Stockum has published numerous articles in the area of law, science and creative writing.

STEVEN PAUL EILERS

Steven P. Eilers was born in Louisville, Kentucky in 1947. He is a self-taught artist and illustrator with works in the American Folk Art Museum in Atlanta, Georgia. His paintings have hung in galleries in San Francisco and Hilton Head. In 1973, Steven opened the Eilers-Hewett Graphic Design Studio.

He retired from Eilers-Hewett in 1990, living for a few years on a sailboat on Little Pine Island in Florida doing illustrations. In 1998, Steven moved into the Clifton area of Louisville, KY where he still lives.

He plays guitar and drums and for many years he performed with the legendary Louisville musician, Tim Krekel. Steven writes music and continues to perform in a number of local bands.

From author Ronald R. Van Stockum Jr.
The Wonder of Natural Life in Kentucky
by
Ronald R. Van Stockum, Jr.
Illustrations by
Steven P. Eilers
Available at Amazon and Kindle; and rvs@vanstockum.com

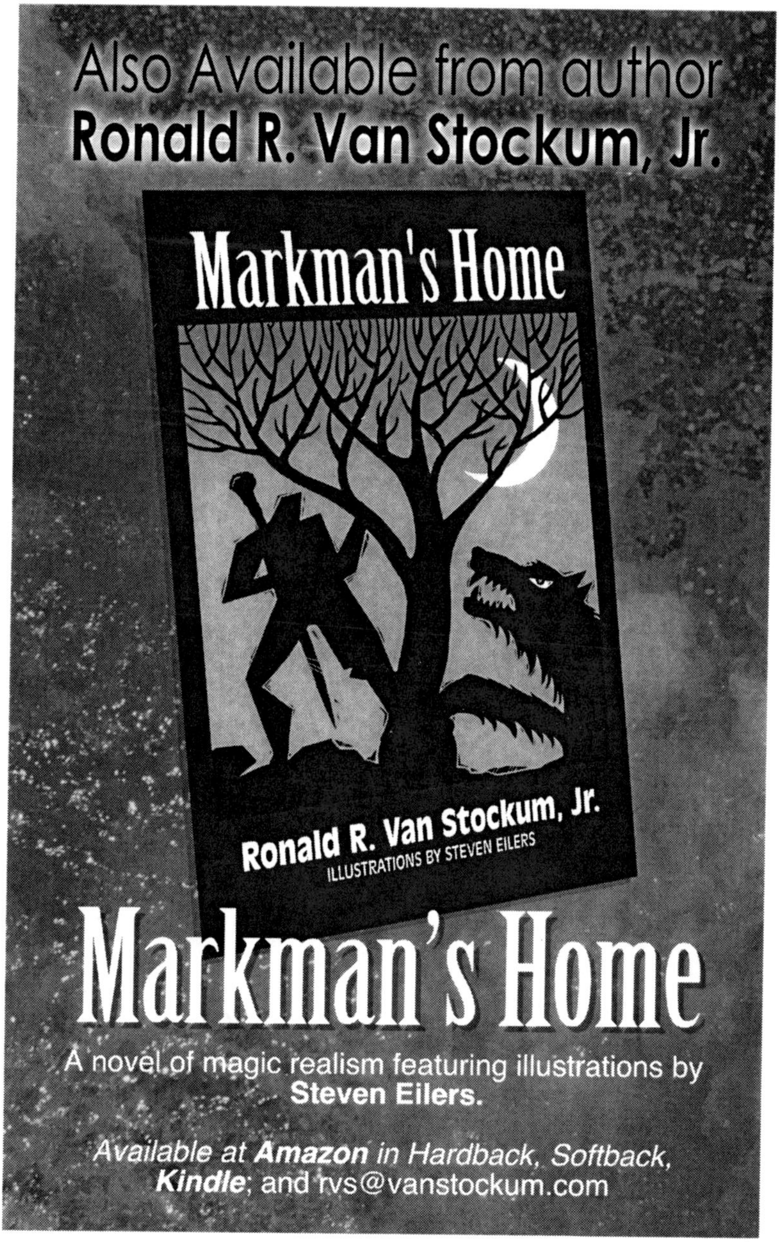
Also Available from author
Ronald R. Van Stockum, Jr.
Markman's Home
Ronald R. Van Stockum, Jr.
ILLUSTRATIONS BY STEVEN EILERS
Markman's Home
A novel of magic realism featuring illustrations by Steven Eilers.
Available at Amazon in Hardback, Softback, Kindle; and rvs@vanstockum.com

From author
Ronald R. Van Stockum, Jr.
MARKMAN'S CAVE
A novel of magic realism.
Illustrations by Steven P. Eilers
Available at Amazon and Kindle; and
rvs@vanstockum.com

Also Available from author
Ronald R. Van Stockum, Jr.

Ollie Deaton &
TheBreathitt County
Silver Mine

A short story featuring
illustrations by **Steven Eilers.**

*Available at **Amazon** and **Kindle***; and
rvs@vanstockum.com

SIDE STEP
RED HOOK SALOON
COLD-BEER
From author
Ronald R. Van Stockum, Jr.
Illustations by Steven P. Eilers
A short story featuring illustrations by Steven Eilers.
Available at Amazon and Kindle; and rvs@vanstockum.com

RONALD R. VAN STOCKUM, JR.
Cleopatra's Needle
ILLUSTRATIONS BY
STEVEN P. EILERS
A short story featuring illustrations by Steven Eilers.
Available at Amazon and Kindle; and rvs@vanstockum.com

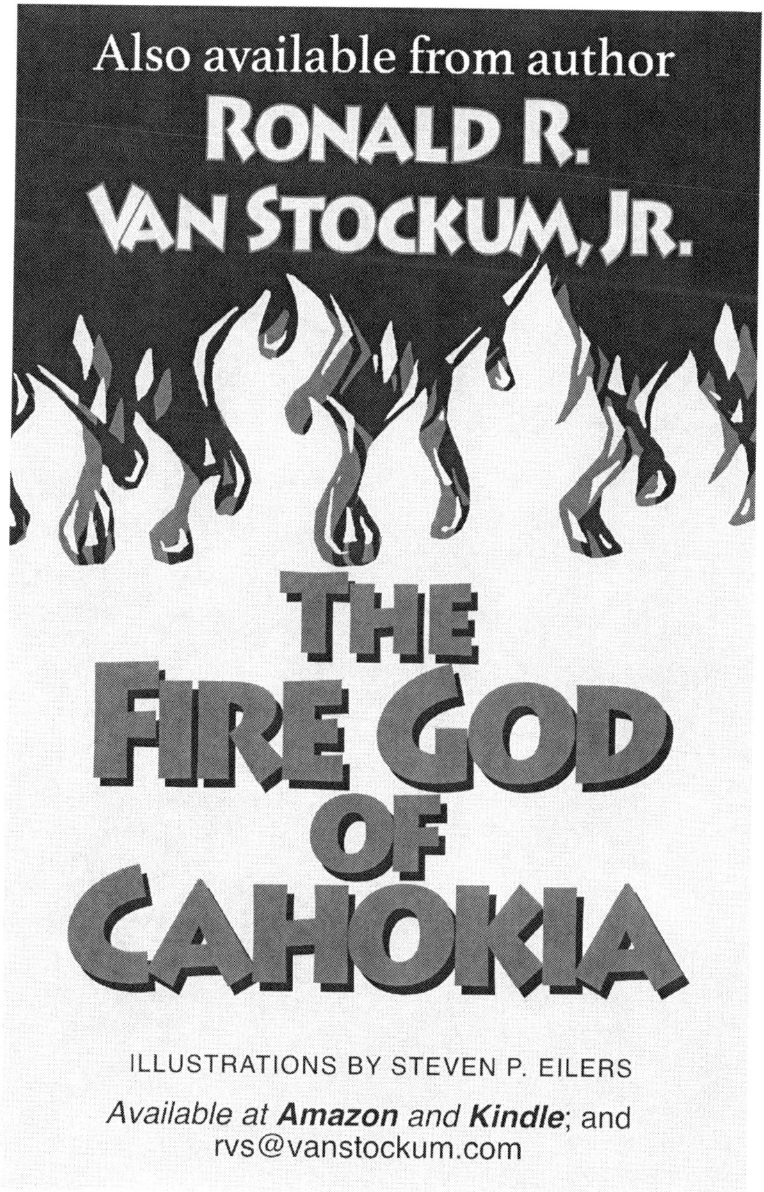
Also available from author
RONALD R.
VAN STOCKUM, JR.
THE
FIRE GOD
OF
CAHOKIA
ILLUSTRATIONS BY STEVEN P. EILERS
Available at Amazon and Kindle; and
rvs@vanstockum.com

CPSIA information can be obtained
at www.ICGtesting.com
Printed in the USA
FFOW01n0631160517
35609FF